Mapping and naming the Moon

A HISTORY OF LUNAR CARTOGRAPHY AND NOMENCLATURE

Almost 30 years after the Apollo missions, 'Tranquillity Base', 'Hadley Rille', or 'Taurus-Littrow' are names still resonant with the enormous achievements represented by the lunar landings. But how did these places get their names? Who named Copernicus crater? Where did all those names on lunar maps come from, and what stimulated the choice of names?

Ewen Whitaker traces the origins and evolution of the present-day systems for naming lunar features, such as craters, mountains, valleys and dark spots. The connections between the prehistoric and historic names, and today's gazetteer are clearly described. Beautiful lunar maps spanning four centuries of progress wonderfully illustrate the unfolding of our ability to map the Moon. Rare, early, photographs add to the sense of history. Comprehensive appendices and the bibliography make this delightful book a work of lasting reference and scholarship.

Ewen Whitaker worked as an astronomer at the Royal Greenwich Observatory (Greenwich and Herstmonceux), Yerkes Observatory (Wisconsin), and the Lunar and Planetary Laboratory (University of Arizona). He is a Member of the International Astronomical Union, and works with its task group on lunar nomenclature. In 1982 the British Astronomical Association awarded him it highest honour, the Walter Goodacre Medal. A participant of several NASA missions, Whitaker located the landing position of *Surveyor 3* which enabled *Apollo 12* astronauts to land alongside it and return parts to Earth. Ewen Whitaker lives in retirement in Tucson, Arizona.

Mapping and naming the Moon

A HISTORY OF LUNAR CARTOGRAPHY AND NOMENCLATURE

EWEN A. WHITAKER

CAMBRIDGE
UNIVERSITY PRESS

PUBLISHED BY THE PRESS SYNDICATE OF THE UNIVERSITY OF CAMBRIDGE
The Pitt Building, Trumpington Street, Cambridge, United Kingdom

CAMBRIDGE UNIVERSITY PRESS
The Edinburgh Building, Cambridge CB2 2RU, UK
40 West 20th Street, New York NY 10011–4211, USA
477 Williamstown Road, Port Melbourne, VIC 3207, Australia
Ruiz de Alarcón 13, 28014 Madrid, Spain
Dock House, The Waterfront, Cape Town 8001, South Africa

http://www.cambridge.org

First published 1999
First paperback edition 2003

Typeset in Swift EF 9.25/14pt, in QuarkXPress™   [SE]

A catalogue record for this book is available from the British Library

Library of Congress Cataloguing in Publication data

Whitaker, Ewen A. (Ewen Adair)
Mapping and naming the moon : a history of lunar cartography and
nomenclature / Ewen A. Whitaker.
     p.   cm.
Includes bibliographical references and index.
ISBN 0 521 62248 4 hardback
1. Moon–Maps–History.   2. Moon–Nomenclature–History.
3. Moon–Gazeteers.   I. Title.
QB595.W48   1999
912.99´1–dc21   98-39475 CIP

ISBN 0 521 62248 4 hardback
ISBN 0 521 54414 9 paperback

CONTENTS

LIST OF ILLUSTRATIONS

PICTURE CREDITS

ACIC/ Pic du Midi; Frontispiece, 23, 35(b)

Royal Greenwich Observatory Library; 8, 16, 21d, 49, 53, 54, 55, 56(a), 58, 59, 60, 61, 95

British Library; 13, 36

Earl of Egremont and Leconfield; 14

Observatoire de Paris; 17, 18 (left), 47(b)

Royal Astronomical Society Library; 22, 29, 30, 32, 33, 34, 35(a), 38, 39, 40, 45, 46, 48, 49, 50, 51(a), 52, 56(b), 57, 94

Cabinet des Estampes, Bibliothèque Nationale, Brussels; 24 (right & bottom)

Observatorio de Marina, Cadiz; 24(left), 37, 42

Crawford Collection, Royal Observatory, Edinburgh; 26

Université de Strasbourg; 27

University of Arizona; 31

Sid Richardson Library, University of Texas; 47(a)

Lunar and Planetary Lab., University of Arizona; 51(b), 71, 79(c)

George Philip and Son; 93

PREFACE

How does one present the subject of lunar nomenclature – which is really no more than a feature identification device for anyone who works on the research, cartography, or mere observation of the Moon's surface – to make it interesting to the general reader? After all, this is material that rightly belongs in the reference section of specialist libraries. To be sure, its history could perhaps provide some facts of academic interest, but such an account would be all too likely to follow the format of a typical paper destined for a scientific journal, beset with references, footnotes, and similar distractions.

In an attempt to make the story more readable, I have given only very general external references in the text, and there are no footnotes. Except for three short lists of feature names, the text is free of tables of nomenclature. That material, such as details of the three pioneer nomenclatures of the mid-seventeenth century, plus the major additions made by later selenographers and committees, is to be found in the appendixes following the text. A selective bibliography of relevant literature is also provided.

EWEN A. WHITAKER
Tucson, Arizona

INTRODUCTION

'Tranquility Base here – the Eagle has landed!'. Millions of people around the world heard these, the first words ever transmitted from the surface of the Moon. But why 'Tranquility' – a name that had no obvious connection with the *Apollo 11* mission? The various commentators explained that the spacecraft had landed in the 'Sea of Tranquility'. So where was the water? The pictures transmitted by the TV camera showed nothing but a level, dry, pock-marked dusty desert.

Meanwhile, transmissions from the orbiting Command and Service module described the scenery passing below, mentioning such features as Langrenus crater, the Ariadaeus Rille, etc. Who on Earth was Langrenus? Who put that name on that particular crater, and when? Likewise, who was Ariadaeus, and what is a Rille, and who coined that word for whatever type of feature it is? Subsequent missions regaled us with further mysterious appellations – Apennine Mountains (don't they run down the centre of Italy?) – Mount Hadley Delta (not a river delta, surely?) – Silver Spur (sounds like a cocktail bar) – Taurus-Littrow (some connection with the Sign of the Zodiac?) – the list goes on . . .

The provenance of such names piqued the curiosity of many people at the time. What arcane tomes and maps were consulted to provide such esoteric data, with some names sounding quite ancient while others seemed to be of recent vintage? Who nowadays produced Moon maps, and when were the first ones drawn; who decided what names would be used to identify the various features of the surface? Were there rules to be observed, and was international agreement necessary? Were all the names in English, or did other countries use translations into their own languages? The whole subject was a closed book to everyone except a very few of us who had been involved and interested in it from the early 1950s – several years before the advent of the Space Age with its sudden focus on the Moon.

Little did we know then that the subject of lunar nomenclature was destined to become one of major importance for some two decades, involving astronomers, geophysicists, astro-geologists, planetary cartographers, the Board of Geographical Names, national and international committees, all involved in a seemingly endless interchange of correspondence, with work sessions occasionally spiced up by personal vendettas and acrimony triggered by obvious evidence of some shady politics and unilateral moves.

However, this was not the first time that the subject had ruffled some astronomical feathers – by the end of the nineteenth century, lack of coordination between lunar cartographers had led to so many instances of a name designating totally different features on different maps that an international committee was convened to clear up the chaos. Even as far back as the middle of the seventeenth century, one selenographer claimed the idea of naming the lunar features as his own, even though a map with no fewer than 325 names had been published two years earlier!

The story of the naming of the various features of the lunar surface is, of course, inextricably bound up with the whole subject of lunar imagery and cartography, although its roots may well antedate the earliest known images and maps of the Moon by a long period. Similar situations hold for the Earth and the heavens, where rivers, lakes, mountains etc. in the former, and constellations and stars in the latter, had received names long before the earliest extant maps of these subjects had made their appearance.

The account that follows is divided, somewhat arbitrarily I must confess, into four parts or eras. The first part focuses on the origins and evolution of the naming of the lunar surface features, from some speculative prehistoric names up to the situation in 1651, when Giovanni Riccioli published a Moon map replete with names, most of which have remained unchanged on their allocated places to this day.

The second part (era) takes us through the period sometimes referred to as 'the long night of selenography', even though some major advances were made then, up to 1837, when the whole subject of lunar cartography and observation was placed on a firm scientific basis by Beer and Mädler.

Both parts are illustrated with reproductions of a large selection of Moon images and maps whether or not they provide any nomenclature. The earlier examples have a fascination all their own, with their sometimes distorted outlines, conjectural details, corner cartouches, and other devices that add greatly to their appeal. In this they parallel to some extent the terrestrial maps of the sixteenth and seventeenth centuries, although to the best of my knowledge, all but one lack the attractive colouring of these. I have added notes on the techniques and conventions used in the preparation of the maps, critiques of their accuracy, and other points of interest that have usually escaped comment elsewhere.

The third era deals with post-Mädler lunar mapping, especially with the nomenclatural nightmares that arose as certain features were often independently given different names or designations by mappers, and how this dilemma took 30 years to fix through an internationally approved map and catalogue.

The fourth era documents the numerous changes and additions that have occurred since the 1935 publication date of those items, due in part to one astronomer's desire to study the Moon as a key to probing its origin and that of the other planets and satellites of the Solar System, but mostly to the acquisition of high-resolution imagery of virtually the entire lunar surface by spacecraft.

To record the full story of the problems, intrigues, politics, acrimony, nonsensical ideas and schemes, and the resulting endless meetings and correspondence that occurred from about 1967 to about 1977 would require a separate book. I have tried to limit the story to the main points that affected nomenclature during that period. The account ends with the comparative 'Tranquillitas' and 'Serenitas' of the mid-1990s.

ACKNOWLEDGEMENTS

As one whose interest in the Moon was first stimulated almost half a century ago, I have been a keen collector of lunar maps, texts, photographs etc. from that time – not with a view to just amassing a collection, but rather to having at hand materials that pertained to the history of lunar observations in general and cartography in particular. I am a firm believer in the idea that if one is to understand a subject well, it is necessary to learn about its roots and subsequent progress.

My efforts at collecting these materials were greatly simplified from the beginning by having direct access to the excellent library at the Royal Observatory, Greenwich, and then Herstmonceux, and by using their copying and darkroom facilities. Being the only 'lunar nut' in that institution, several outdated maps and photographs were eagerly dumped on me during the move from Greenwich to Herstmonceux. As a Fellow of the Royal Astronomical Society, I also had access to their comprehensive library, and was given carte blanche permission by the Council to borrow relevant materials for copying. I extend thanks posthumously to the late Dr Alan Hunter and the then Council for granting this special privilege. My gratitude also to my lunar colleague and friend for many years, D.W. Arthur, for passing on to me his collection of lunar maps and personal notebooks, also books and reprints from S.A. Saunder's effects originally donated to him by the University Observatory, Oxford. I am also grateful to many other kind people who have donated odd maps, photos, books, reprints etc. to that 'lunatic collector' over the past 45 years. My own purchases over this period have added to the total collection, upon which I have largely drawn for the data and illustrations in this book.

I also thank the late Dr W. Rubey for securing a small grant from the Lunar

Science Institute, Houston, to copy further materials in England and Paris immediately before the IAU Congress in Brighton, 1970. For invaluable assistance on that occasion, my thanks go to Joan Perry, then RGO librarian; the late R.E.W. Maddison, RAS librarian; F. Maddison, then Director of the Museum for the History of Science, Oxford; Mme. G. Feuillebois, then Observatoire de Paris librarian; Mrs P. Gill, Archivist for the West Sussex County Council, who had temporary guardianship of Harriot's lunar observing book, and to the Earl of Egremont and Leconfield for allowing this privilege.

My grateful thanks also to A. Orte, Director of the Instituto y Observatorio de Marina, Cadiz, for providing negatives of their historic lunar maps; H. Michel, Société Belge d'Astronomie, for providing negatives of two of Mellan's images; O. van der Vyver, S.J., for providing negatives of early Moon images, his booklet on the correspondence of Van Langren, and other information; the Sid Richardson Library, University of Texas, Austin, for providing a negative of Cassini's 1680 lunar map; and the Crawford Library, Royal Observatory, Edinburgh, for supplying a negative of Van Langren's engraved map.

I must thank Godfrey Sill for translating the Latin inscriptions on Van Langren's map, also Hevelius's lengthy description of his choice of nomenclature; for shorter pieces in Latin, French, and German I have relied on my own memory from schoolboy days in the mid-1930s. Also thanks to Humberto Campins for translating some seventeenth-century Spanish, and to George Coyne for help with some Italian. I am also grateful to Philip Stooke and Scott Montgomery for alerting me to the Irish Stone Age and the Van Eyck images respectively, and for kindly sending me their researches on these.

Grateful thanks also to Lisa Martin, Lunar and Planetary Laboratory librarian, for invaluable assistance with tracking down some journal and book references. Finally I wish to acknowledge the constant encouragement of many friends and colleagues, without which my inborn procrastinatory proclivities, together with the desire to get back to restoring antique clocks, would have left the half-finished manuscript a'moulderin'.

FIRST ERA

FROM PREHISTORIC IMAGES TO ARCHETYPE MAP

CHAPTER 1

....

PRE-TELESCOPIC LUNAR OBSERVATIONS

IMAGES IN THE MOON

Mankind has a natural propensity for seeing the shapes or images of familiar objects in such unlikely media as clouds, rocky outcrops or cliffs, profiles of distant mountains, and so on. This facility includes the night sky, of course, and the natural groupings of the stars have been likened to various animals, personages, inanimate objects etc. since prehistoric times. The pattern of light and dark markings visible on the Moon, notably at or near the full phase, has likewise evoked similar comparisons, resulting in several quite divergent imagined likenesses such as a human face, an old man carrying a bundle of sticks, an old lady spinning, two children carrying a bucket, several different rabbits, and so on. The Chinese, who imagined a rabbit sitting on its haunches pounding rice, perhaps referred to our Mare Tranquillitatis as 'the head of the rabbit', and the bright area surrounding our Copernicus crater as 'the pile of rice' (fig. 1b), while our Mare Serenitatis and Mare Imbrium might have been the 'left eye' and 'right eye' respectively of the full-faced Man in the Moon (fig. 1c) to those cultures that maintained this particular image. If so, those appellations could be considered to constitute a rudimentary, ancient lunar nomenclature.

At least part of this diversity in the images is due to the fact that, although the Moon presents almost exactly the same face to us at all times, the angle at which this face is oriented to our view can vary widely, and the imagination favours upright subjects.

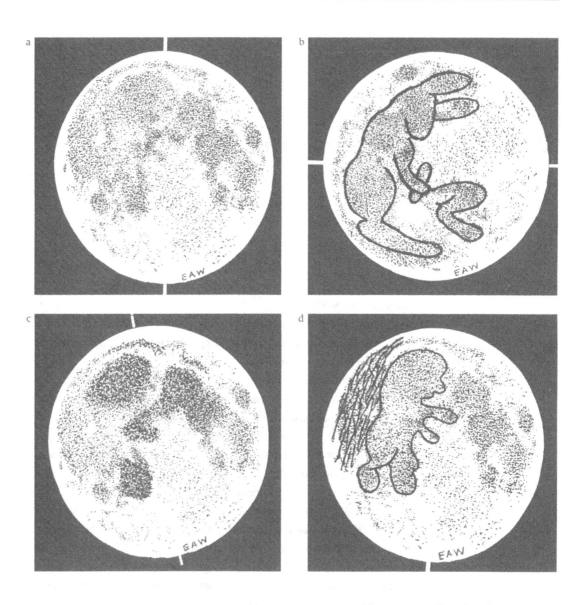

NAKED EYE VIEWING OF THE MOON

The Moon's markings are, of course, fixed relative to its axis, but this axis may be situated at any position angle with respect to the viewer's local vertical. This angle depends upon (a) the Moon's position in its orbit, (b) the positions of its orbital nodes on the ecliptic, (c) its position in the sky, and (d) the viewer's latitude (see fig. 2). Another relevant factor, one that affects the clarity of the markings, is that of the Moon's glare at the full phase. Its dazzling brilliance in a black sky notably lessens the visibility of

Fig. 1. Imagined shapes in the Moon. (a) Naked eye view of full Moon, showing somewhat finer detail than normally seen; (b) the oriental rabbit, pounding a heap of rice; (c) the familiar 'Man in the Moon'; (d) old man carrying a bundle of sticks.

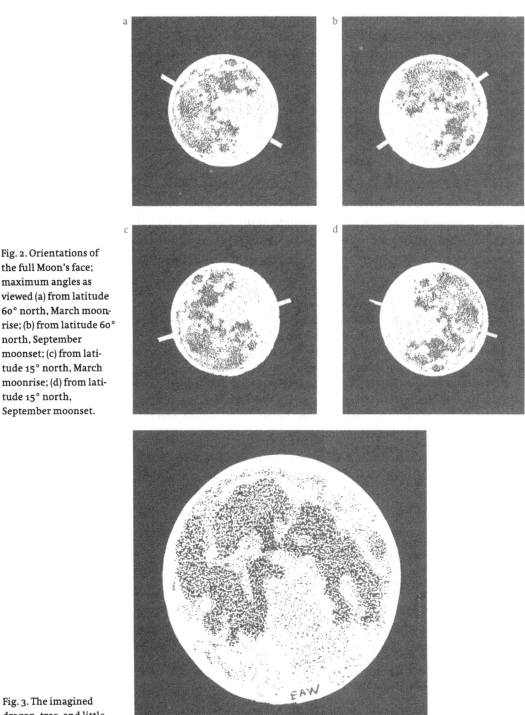

Fig. 2. Orientations of the full Moon's face; maximum angles as viewed (a) from latitude 60° north, March moonrise; (b) from latitude 60° north, September moonset; (c) from latitude 15° north, March moonrise; (d) from latitude 15° north, September moonset.

Fig. 3. The imagined dragon, tree, and little man of Albertus Magnus.

those markings; they appear to be more prominent in a twilit sky, where the muted background brightness eliminates the dazzle. Since the full or nearly full Moon rises near sunset and sets near sunrise, it follows that the full array of markings is most noticeable when the Moon is low in the eastern or western sky, and thus oriented at a considerable angle, clockwise or anticlockwise, to its average orientation when crossing the meridian. The Chinese rabbit is an excellent example of a 'springtime moonrise' image. We will remark upon this selection effect a little further in a later section.

Most of the descriptions of images seen in the Moon are closely entwined with various legends and folklore, which suggests ancient origins. A more recent example of these imagined shapes is the group of **the dragon, the tree**, and **the little man** (fig. 3) which may be traced back to Albertus Magnus (Saint Albert of Böllstadt, c. 1193–1280). Here is how he describes his impressions.

> It seems to me that this shading is to be found on the eastern part of the Moon towards the lower limb, having the figure of a dragon which has its head towards the west and its tail towards the east with respect to the lower limb; the tip of the tail is not sharp, but large like a leaf with three contiguous circular segments. Above the dragon's back stands the figure of a tree from which the branches proceed obliquely from the centre of the trunk towards the lower eastern part of the Moon, and leaning against the oblique part of the trunk with head and elbows there is a man whose legs descend from the upper part of the Moon towards the western part.

Some 350 or so years later these became known to Shakespeare in the slightly changed form of **the dog, the bush**, and **the man** (fig. 4); as we will see later, Harriot soon afterward referred to **'the body of the feyn'd man'** in describing one of his telescopic observations of the Moon, while Gassendi included **Homuncio**, i.e. the little man, in his nomenclature scheme of the 1630s.

GEOGRAPHICAL FEATURES IN THE MOON

Two other names of a quite different nature that seem to have survived from ancient times to the first half of the seventeenth century are **Caspia** and **Penetralia Hecates** (Kaspion and 'Ekatis Mykhos in Greek). They are mentioned by Plutarch (c. 46–120 AD) in his *De Facie in Orbe Lunae [Concerning the Face in the Disk of the Moon]* thus:

> Just as our Earth has deep and great gulfs, one of which flows to us through the Pillars of Hercules, another outside is that of the Caspian Sea, and also that of the Red Sea [now the Arabian Sea], so on the Moon there

are hollows and deeps; they call the greatest of these the **Shrine of Hecate**, where the souls endure or exact retribution for all the things which they have suffered or done ever since they became spirits. Two of them are long, through which the souls pass, first to the parts of the Moon which are turned towards Heaven, then back to the side next to Earth. The parts of the Moon towards Heaven are called '**The Elysian Plain**', those towards Earth '**The Plain of Persephone** [Proserpina in Latin] **Antichthon**' [my emphasis].

Plutarch was not, strictly speaking, a philosopher, but was in fact a well-educated and very able Greek biographer, chronicler and publicist. Here, he was reporting the current ideas of those who followed the teachings of Thales, Democritus, Anaximenes and others who believed that the Moon was essentially an Earthlike body. Hecate was identified with Persephone, Diana and Luna in ancient mythology, whereby the use of two of those names here.

Although Plutarch does not exemplify the **Caspian Sea** as anything more than an illustration of the type of feature that might exist on the Moon, at some time between the availability of his writings and the beginnings of the seventeenth century, some person or persons unknown apparently gave that name to the dark spot that is now known as Mare Crisium, since Harriot, Gassendi and Van Langren independently used that appellation for it. Possibly it was so named because it occupies roughly the same position with respect to the Moon's face that the Caspian Sea does with respect to a map of Europe, N. Africa, and the Middle East. As Harriot's manuscript (MS) notes were never seen by Gassendi or Van Langren, it seems that this designation was rather widely known. However, of perhaps more interest is the fact that this appears to be the first instance in which a dark lunar spot had received the name of a terrestrial sea. The **Shrine of Hecate** probably refers to our Mare Imbrium, which from our Earthly viewpoint is the largest regular-shaped dark area unbroken by bright patches. The two long gulfs are Mare Tranquillitatis plus Mare Fecunditatis to the north, and Mare Tranquillitatis plus Sinus Asperitatis and Mare Nectaris to the south (fig. 5).

While all this may have been the major influence in subsequently establishing the use of watery designations for the lunar dark spots, a misreading of Galileo's words on the subject probably ensured the perpetuation of this inappropriate categorization. However, we are getting ahead of ourselves somewhat. We need to find out whether pictorial representations of the Moon's face accompanied these rather elementary namings of the spots.

 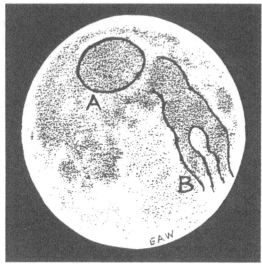

THE FIRST KNOWN MOON DRAWINGS

The many diverse images seen in the Moon over the past millennia bear witness to the degree of scrutiny undergone by that body, and yet it seems that no really serious representations of the pattern of Moon spots have come down to us from antiquity. A recent study by Stooke of concentric arc designs on some Neolithic stones in Ireland makes a good case for their being images of the general pattern of the lunar maria, but they are mostly quite stylized in appearance (fig. 6). My own recent studies tend to show that representations of the Moon found on various artifacts from the Mesoamerican and nearby regions, although assumed to be allegorical, display images of a rabbit that bear remarkable resemblances to the actual Moon markings as viewed in the different orientations explained earlier (fig. 7).

Depictions of the Moon's crescent shape are not too uncommon, being found as petroglyphs, wall paintings, etc., but medieval MSS and early printed books appear to have nothing beyond allegorical images depicting a human face, usually in the Earthlit portion of an otherwise crescent Moon (fig. 8). Since the Sun is usually given a quite similar face, it cannot be claimed that we are seeing the Man in the Moon, although some images of the Earthlit Moon in stained glass church windows could possibly convince us otherwise (fig. 9)!

As might perhaps have been expected, the multi-talented Leonardo da Vinci (1452–1519) turned his attention to the Moon on many occasions.

Fig. 4. (above left) Shakespeare's dog, bush and little man.

Fig. 5. (above right) Plutarch's 'Shrine of Hecate' (A) and long gulfs (B).

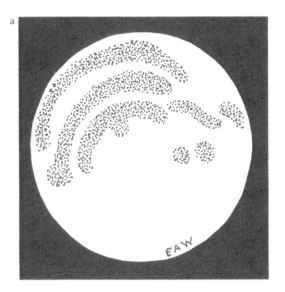

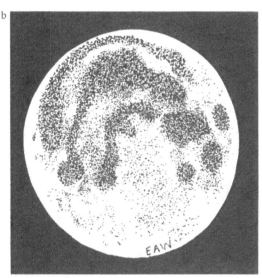

Fig. 6. Neolithic stone markings (a) (from Stooke), with comparison (b).

Viewing the markings with a scientific rather than an imaginative eye, he refers to them merely as 'spots' (macchie), without singling out any specific parts. He notes in one place that '... the details of the spots of the Moon ...' often show '... great variation ...', which, he writes, he has '... proved by drawing them.' At first sight, this observation seems to be refuted by another observation in which he shows that, contrary to a prevailing opinion at that time, the Moon cannot be a convex mirror reflecting the Earth's continents and seas. Here he notes that '... the spots on the Moon, as they are seen at full Moon, never vary in the course of its motion over our hemisphere.' This apparent contradiction may possibly be explained by the fact that the markings so noticeable at the full phase are almost lost when they are situated near the terminator (the sunrise or sunset line) at other phases. More about this phenomenon later.

Until very recently it was thought that Leonardo was the first to draw a true-to-life image of the Moon's markings, but Montgomery has tracked down three small images that are included in paintings by Jan van Eyck, the best of these images being of a waning gibbous Moon hanging low in a western sky some time after sunrise (fig. 10), in a painting of the Crucifixion.

Unfortunately Leonardo's sketch is only one half of a complete image, the other half either being lost or yet to be discovered (fig. 11). It has a diameter of about 7 inches, and was drawn in black and white chalk in about 1513. The other drawings of the Moon that he mentions are likewise lost or undiscovered, apart from three or four very rough, small sketches.

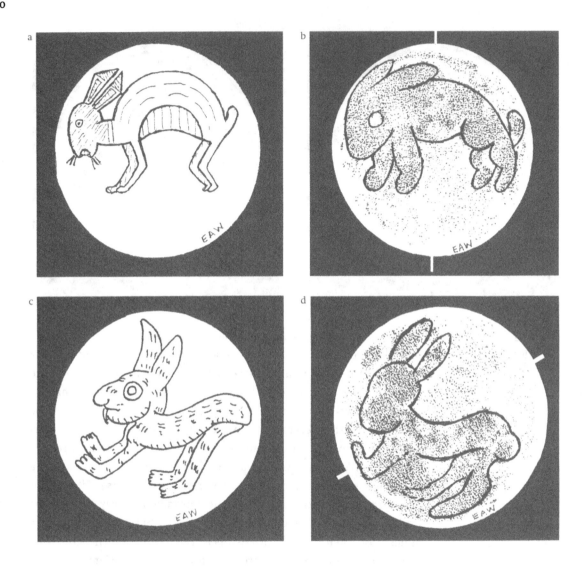

A MOON DRAWING WITH NAMES

The only other currently known pre-telescopic image of the Moon's spotty face is the pen-and-ink sketch of the full Moon made by William Gilbert (1540–1603), physician to Queen Elizabeth I and author of the famous *De Magnete* [*Concerning the Lodestone*]. This was drawn in about 1600 and is included in the original MS of his book *De Mundo Nostro Sublunari Philosophia Nova* [*New Philosophy Concerning our Sublunar World*]. In this he regretted that no images of the Moon's face had come down from antiquity, preventing the possibility of detecting any major changes in the markings. Contrary to the idea that the dark areas represented seas, he

Fig. 7. More imagined rabbits in the Moon. (a) Stylized Mimbres rabbit with (b) comparison; (c) Mexican rabbit with (d) comparison; (e) two Mexican rabbits and a Mayan, with tentative comparisons in the waxing and waning phases.

Fig. 7. *(cont.)* e

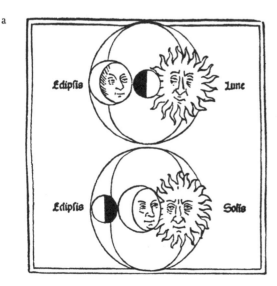

Fig. 8. (below) Typical
early representations of
the Moon and Sun.

a

b

Eclipſis   Lune

Eclipſis   Solis

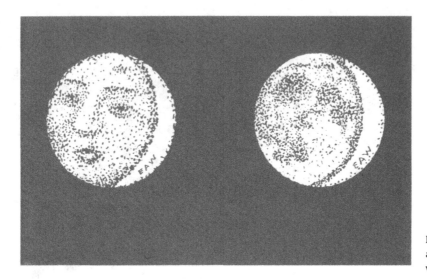

Fig. 9. The Moon's face in a church window (Erfurt), with comparison.

Fig. 10. Facsimile of image of the waning Moon in a van Eyck painting.

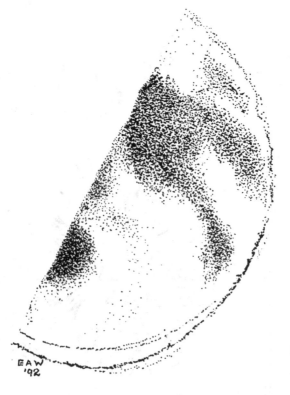

Fig. 11. Facsimile of half a full-Moon image by Leonardo da Vinci.

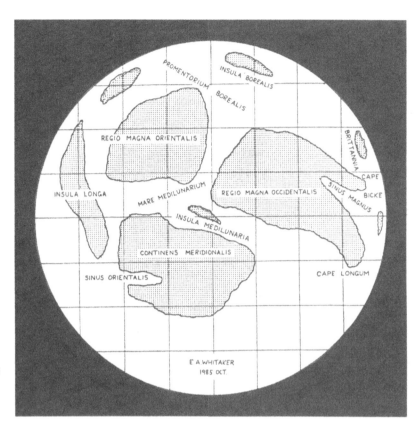

Fig. 12. Facsimile of full Moon 'map' by William Gilbert. the original, drawn from his naked eye observations, is unsuitable for reproduction at small scale.

thought (as did Leonardo da Vinci and, at first, Kepler) that the opposite was true. This misconception probably stemmed from a lack of appreciation of the difference between specular reflection and diffuse scattering of light, as well as the quite different effects produced either by clear sunshine or by a veiled but bright sky on a terrestrial scene with land and water.

Figure 12 is based on this 11-inch-diameter sketch, but the lettering has been enlarged and capitalized, and the dark areas shaded for better clarity. The original is interesting in that Gilbert used an 8 × 8 square grid to better position the features, but mislocated some outlines one square too far north before realizing his error. Figure 13 illustrates this, and also points to a quite different method of drawing the Moon's face – less artistic than Leonardo's but perfectly valid nevertheless. It qualifies as a map rather than an image.

Gilbert includes 13 names in all on this map, which are listed here with translations and modern equivalents. General terms used are **Mare**,

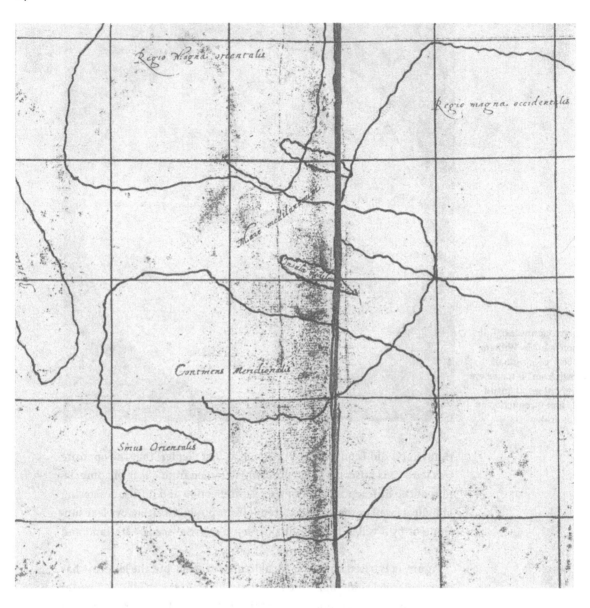

Sinus, Continens, Regio, Insula, Promentorium [*sic*], and **C**(aput), which translate as Sea, Bay, Continent, Region, Island, Promontory, and Cape or Headland. Gilbert's map and its accompanying text were not published until 1651, by which time three major nomenclature schemes had made their appearance on maps prepared from observations made with telescopes; hence it had absolutely no impact on lunar nomenclature or cartography.

Fig. 13. Central part of Gilbert's original MS map, showing his placement error of one square. His very thin lines are enhanced here.

GILBERT'S NOMENCLATURE

Gilbert	Translation	Modern
Brittannia	Britain	M. Crisium
C. Bicke		anon.
C. Longum	Long Cape	anon.
Continens Meridionalis	Southern Continent	M. Nubium etc.*
Insula Borealis	Northern Island	M. Frigoris (part of)
Insula Longa	Long Island	O. Procellarum (part of)
Insula Medilunaria	Middlemoon Island	S. Medii
Mare Medilunarium	Middlemoon Sea	anon.
Promentorium Borealis	Northern Promontory	anon.
Regio Magna Occidentalis	Large Western Region	M. Serenitatis etc.**
Regio Magna Orientalis	Large Eastern Region	M. Imbrium
Sinus Magnus	Large Bay	anon.
Sinus Orientalis	Eastern Bay	anon.

* Includes M. Cognitum, M. Insularum, M. Humorum.
** Includes M. Tranquillitatis, M. Fecunditatis (northern part), M. Nectaris, and
S. Asperitatis.

....

CHAPTER 2

EARLY TELESCOPIC OBSERVATIONS
OF THE MOON

THOMAS HARRIOT

Although Galileo is usually credited as being the first to examine the Moon through a telescope, Thomas Harriot (*c.* 1560–1621), an astute scientist and mathematician who is perhaps better known for his description of an expedition with Sir Walter Raleigh to Virginia in 1585, and for the introduction of the 'greater than' and 'less than' signs in mathematics, actually made the first known telescopic sketch of the Moon – on 5 August 1609 (Gregorian), four months before Galileo's first observation. He twice refers to '**the Caspian**' (our Mare Crisium, as already noted) in notes accompanying lunar sketches that were made towards the end of 1610, and also twice mentions the '**body of the man**' in the Moon, each time clearly referring to our Mare Tranquillitatis, confirming that at least these two names were extant at that time.

Harriot was also the first to produce, in about 1611, what may be described as a telescopic map of the Moon, although technically it is a pen-and-ink drawing, 6 inches in diameter, of the full Moon with letter and number annotations (fig. 14). Comparison with the full Moon photograph in the frontispiece shows that while Harriot was certainly no great artist, he made an honest attempt to portray the main features, and included a number of readily identifiable bright surface spots. Notes accompanying the drawing have several entries such as: '1, 23, 15, a right lyne', and '1, e, a, equilaterall', written in an older hand, showing that he was trying to position points with some degree of precision. His other 15 sketches, all of phases, are quite poor in comparison (fig. 15). Thus we see that he used these annotations exclusively as positioning guides; however, they could

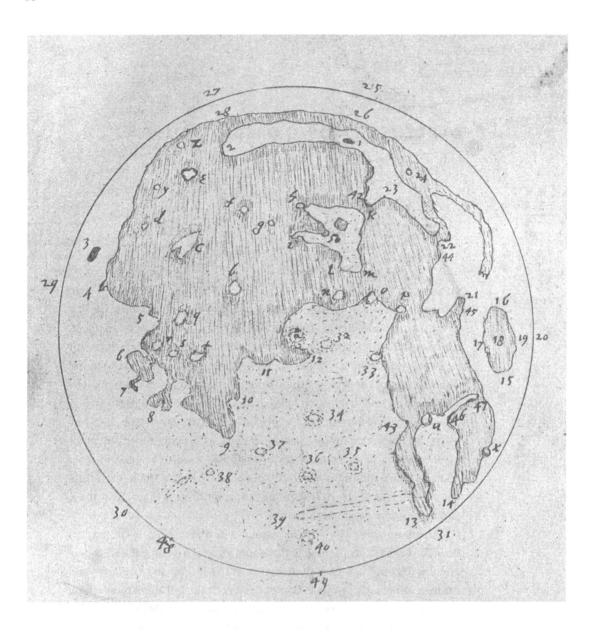

be construed as constituting a simple form of nomenclature akin to Bayer letters for the stars. As we will see, five later observers (Scheiner, Biancani, Borri, Fontana, and Rheita – see figs. 21, 22, and 29) also used a few letter designations on their published images, but each scheme was different and was again intended only as a simple guide for the accompanying text. Harriot's map remained virtually unknown until its publication in 1965.

Fig. 14. Thomas Harriot's MS map of full Moon, showing his letters and numbers used for identification purposes.

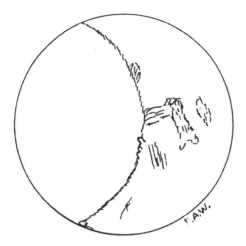

Fig. 15. Facsimiles of two of Harriot's sketches of lunar phases.

GALILEO GALILEI

Galileo (1564–1642), who was the first to publish – in his book *Sidereus Nuncius* [*Sidereal Message*] in 1610 – sketches and descriptions of the lunar surface made with a telescope, used no form of nomenclature at all, merely referring to the large dark areas as the 'great or ancient spots' [maculae magnae sive antiquae]. His observations, however, settled the ancient controversy over whether the Moon was basically an Earthlike body or something far more exotic such as a crystalline sphere, a body of condensed fire, and so on. The brighter regions were, he noted:

> . . . full of inequalities, uneven, full of hollows and protuberances, just like the surface of the Earth itself which is varied everywhere by lofty mountains and deep valleys.

With regard to the large dark spots, he writes:

> . . . (these) are not seen to be at all similarly broken, or full of depressions and prominences, but rather to be even and uniform; for only here and there some spaces, rather brighter than the rest, crop up; so that if anyone wishes to revive the old opinion of the Pythagoreans, that the Moon is another Earth, so to say, the brighter portion may very fitly represent the surface of the land ['terra'], and the darker the expanse of water ['aqua']. Indeed, I have never doubted that if the sphere of the Earth were seen from a distance, when flooded with the Sun's rays, that part of the surface which is land would present itself to view as brighter, and that which is water as darker in comparison.

He further observed that 'the great spots in the Moon are seen to be more depressed than the brighter tracts', as is the case with terrestrial seas and lands.

'MARIA' AND 'TERRAE' ON THE MOON

By asserting that water would appear darker than land when viewed from afar, Galileo was in agreement with Plutarch and in direct opposition to Kepler (1571–1630) who, in his *Dioptrice* (1604), wrote that he believed the brighter parts of the Moon to be seas (maria) and the darker, spotty areas to be lands (terrae), continents, and islands. Kepler knew of Plutarch's writings on this subject, having translated them from the original Greek into Latin. However, on reading Galileo's pronouncements, Kepler reconsidered his opinion, which had been based on an observation of land and river from a mountain. In his *Dissertatio cum Nuncio Sidereo* [*Conversation with the 'Sidereal Message'*], in 1610, Kepler writes '. . . do maculas esse maria, do lucidas partes esse terram.' [I admit that the spots are seas, I admit that the bright areas are land.] – bold assertions indeed.

We note with interest that Galileo, unlike Kepler, chose his words carefully, and nowhere does he say that he himself considers the dark spots to be stretches of water, and he never uses the words 'mare' or 'maria'. Indeed, in a letter dated 28 February 1616 to Girolamo Muti, he writes, 'I said then and I say now that I do not believe that the body of the Moon is composed of earth and water.' He reiterates this view at greater length in his *Dialogue on the Two Chief World Systems*, noting that factors other than land and water could well be the cause of the brightness contrasts.

Galileo's *Sidereus Nuncius* had a wide circulation, and most of its readers probably construed the description of the dark spots as meaning that they were more likely to be seas than otherwise. The wording certainly seems to favour this interpretation, even though Galileo leaves the option to the reader ('. . . if anyone wishes . . . the brighter portion **may** . . .') [my emphasis]. To those who knew of Plutarch's writings, or of the Caspian Sea appellation, Galileo's assertions undoubtedly added considerable weight to the interpretation, and Kepler's turnaround agreement and use of the words 'mare' and 'terra' as representing actual sea and land on the Moon possibly clinched it. In any case, the idea was apparently sufficiently widespread that the three pioneer selenographers of the mid-1600s – Van Langren, Hevelius, and Riccioli – all named the dark areas as seas, bays, lakes, and one ocean, connotations that have persisted to this day.

GALILEO'S DRAWINGS OF THE MOON

These have received much adverse criticism from soon after their first appearance up to very recent times, the remarks becoming more critical as time has passed. Some of this can be attributed to the almost unbeliev-

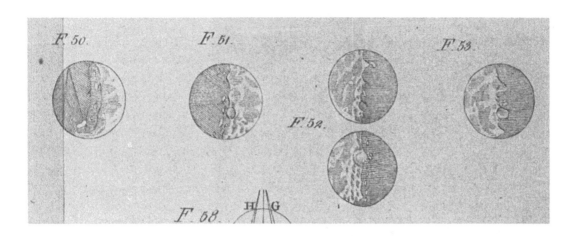

Fig. 16. Grossly degraded and reduced copies of Galileo's engraved images.

able deterioration in image quality and size that occurred in successive editions of Galileo's *Opere* [i.e. *Works*], all of which include *Sidereus Nuncius* (fig. 16), but much stems from perfunctory examinations of the images. The book contains five copperplate imges, all about 8 cm in diameter, of four different phases, one image being repeated to adjoin the relevant text. The printer placed the image of first quarter upside down, which may have caused some confusion when compared with the images of last quarter. The four images are shown properly oriented in fig. 17. Figure 18, in which a slightly enhanced and cleaned up version of the last quarter image is compared with a low resolution photo of the Moon at the same phase, should suffice to muzzle the gainsayers who assert that Galileo's images bear no resemblance to reality.

Apart from a small, rough sketch obviously made from memory and partially written over with calculations, Galileo did not make a drawing of the full Moon, the reason being that he was interested in spreading the news of the Moon's considerable roughness, which is not apparent at the full phase. However, it is possible to mosaic parts from three of his four images and to fill in the poorly illuminated areas, resulting in the 'fake' full-Moon image shown in fig. 19, which compares quite favourably with Harriot's image. Galileo's images are superior to all but one (by Scheiner) that were published during the next quarter of a century.

Among the Galileo manuscripts there exist two pages with seven brown-wash images of different Moon phases, which are probably copies of the original sketches made at the telescope (fig. 20). A careful study of these and the four engravings has led to the determination of the dates on which the observations were made – a matter of much speculation in the

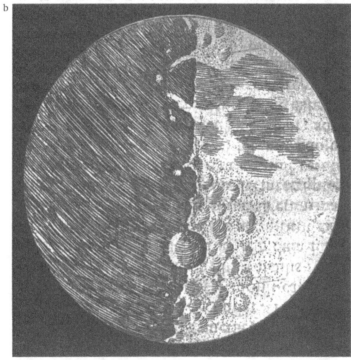

Fig. 17. Galileo's original engraved images, 1610: (a) crescent phase, (b) first quarter, (c) waning gibbous phase, (d) last quarter.

Fig. 17. (*cont.*)

c

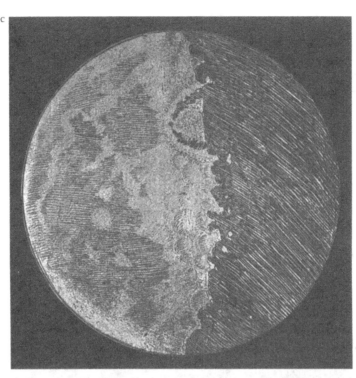

d

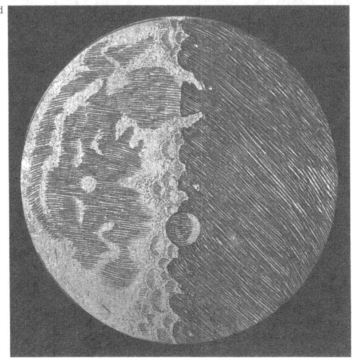

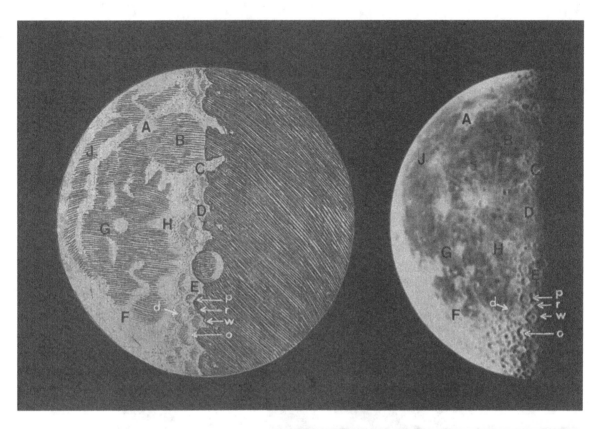

Fig. 18. (above) Galileo's image of last quarter with photo for comparison, showing many correlations.

Fig. 19. (right) 'Fake' full-Moon image made by combining parts of three of Galileo's engravings and filling in the shadowed areas in Mare Serenitatis, Mare Vaporum, and Mare Imbrium (cf. fig. 17).

past. His first observation was made in 1609, on 30 November, some four months after Harriot's first recorded sketch.

SCHEINER, MALAPERT, BIANCANI, BORRI, AND FONTANA

Between the publication of *Sidereus Nuncius* and the three beautiful engravings by Claud Mellan in about 1637, the only original Moon images known to me are those by these five authors (figs. 21 and 22). The Scheiner image of first quarter was published in 1614; the dark areas are reasonably well portrayed, and a few bright spots and recognizable craters are shown. The original has a diameter of 9 cm. The Malapert (1619) and Biancani (1620) sketches have almost no merit; their diameters are each about 5 cm. The Borri image of first quarter (10 cm diameter) also has low accuracy. The writing notes that it was made at Coimbra on 14 (18?) July 1627, the Moon's age being six days. It also says 'Exact telescopic appearance of the waxing Moon' – certainly an over-confident statement!

The Fontana images are somewhat of a puzzle. This Neapolitan lawyer, telescope maker and observer made copperplate engravings of two observations of the Moon, the first being on 31 October 1629, the Moon being about two days past full, and the second on 20 June 1930 when its age was about ten days. Although he apparently never published these images at that time, he may have distributed a small number of copies of either the original drawings or the engravings, since the two appear in several books published in the 1640s and even in later editions over the next decade or two, by which time far superior images and maps by other authors had become widely available. Figure 22 shows these two images (diameter about 10 cm) as later reproduced in his 1646 book *Novae Coelestium Terrestriumque Rerum Observationes [New Observations of Celestial and Terrestrial Things]*. Note that they are inverted with respect to all earlier images, the result of using a positive eyepiece in his telescope rather than a concave (Galilean) lens. More on Fontana later.

PIERRE GASSENDI

One of the first to attempt to provide names for lunar features that were revealed by the telescope was Pierre Gassendi (1591–1655), a professor of mathematics in Paris and a leading astronomical thinker of his day. At that time, a major problem in navigation and terrestrial cartography was the very inaccurate longitude values on world maps, and the great difficulty of trying to determine longitudes at sea. It was well known that the passage of the Earth's shadow across the smaller spots of the lunar disc

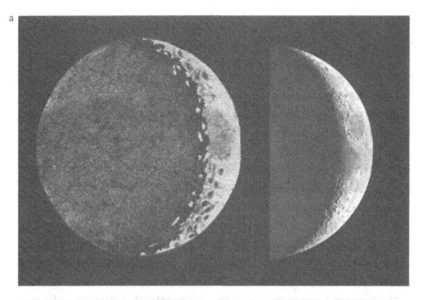

Fig. 20. Four of Galileo's wash images, with comparison photos. The last image shows the 'Earthshine' very well.

during a total lunar eclipse, if simultaneously timed from different locations, could yield improved differential longitudes between those locations.

Encouraged by the success of this method for determining the longitude difference between Aix-en-Provence and Paris after observing the lunar eclipse of 20 January 1628, Gassendi and his friend Nicolas de Peiresc (1580–1637), who discovered the great Orion Nebula in 1610, set in motion a scheme to produce not only an accurate image of the full Moon

Fig. 20. (*cont.*)

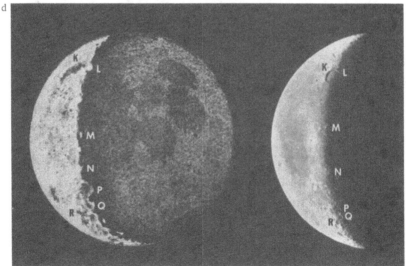

as an indispensable adjunct to these terrestrial longitude determinations, but also an atlas of images at other phases plus a map of the Moon prepared by combining the imagery.

These astronomers were certainly aware of the fact that lunar topography is thrown into prominent relief near the terminator (i.e. the sunrise/sunset line), but that this relief vanishes in regions well removed from the terminator and also at the full phase, when the only markings visible are those caused by differences in reflectivity (albedo) of the

a

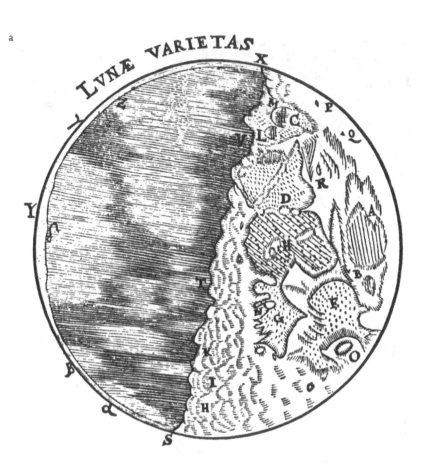

Fig. 21. Images by (a) Christopher Scheiner (1614), (b) Charles Malapert (1619), (c) Giuseppe Biancani (1620), and (d) Christopher Borri (1627).

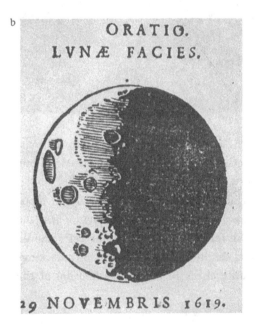

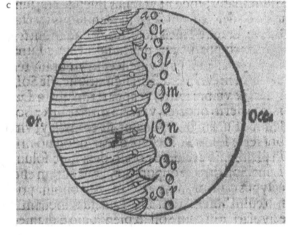

Fig. 21. (*cont.*)

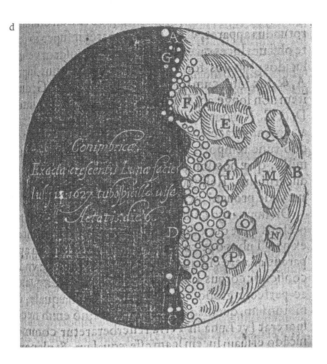

surface materials. The albedo differences, in turn, are scarcely noticeable in regions situated fairly near the terminator, while some (e.g. the crater rays) show no topographic expression under even the lowest angle of illumination (fig. 23). Thus in order to draw up a complete map of the Moon, exhibiting both the albedo markings and the topographic details, accurate drawings made at many different phases are needed.

Gassendi and Peiresc engaged the services of three artists to produce the various phase images, but following some unsatisfactory results, retained one, Claud Mellan, a prominent and very skilled engraver. Excellent copperplate engravings of three different lunar phases, diameter of each about 20 cm, were prepared by this artist and reproduced early in 1637 (fig. 24), apparently in very limited numbers, since they are now extremely rare. Unfortunately, Peiresc died later that year, and the whole project progressed no further.

GASSENDI'S NOMENCLATURE

During the planning stages of this project, Gassendi undoubtedly realized that a map would require a nomenclature scheme, and that names would be preferable to letters or numbers, which cannot readily be committed to

a

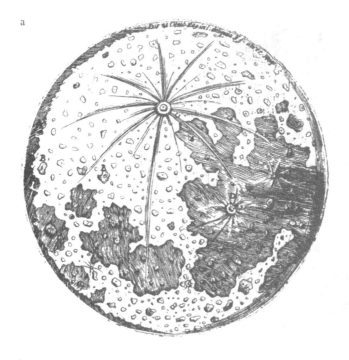

b

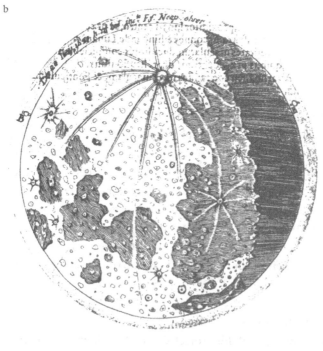

Fig. 22. Francesco Fontana's earliest images, dating from 1629 and 1630; his use of convex eyepiece lenses gave him south-up views. The reversed lettering is bleed-through from the next page.

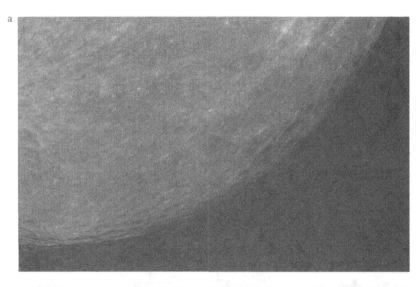

Fig. 23. Photographs of the same lunar area taken with (a) vertical (i.e. full Moon) solar illumination, and (b) slanting (first quarter) illumination, demonstrating that when the topography is well shown, the rays are not seen, and vice versa.

memory and are subject to errors in transcription. With the death of Peiresc and the resulting demise of the map project however, Gassendi's rudimentary nomenclature was never published, and is to be found only in his diaries and letters. Without any accompanying drawings, it is difficult to correlate many of his names with actual features; the list of 14 names with, where possible, modern equivalents (as determined by Humbert) is given below. The only general terms used are **Mare**, **Vallis**, **Rupes**, and **Mons** (Sea, Valley, Cliff or Scarp, and Mount or Mountain).

Fig. 24. Remarkable engravings of three lunar phases by Claud Mellan, dating from 1635–7.

Gassendi	*Modern*
1 **Caspia** (Caspian Sea)	Mare Crisium
2 **Anticaspia** (opposite the Caspian)	Mare Humorum
3 **Homuncio** or **Thersite**	Mare Serenitatis etc.
4 **Umbilicus Lunaris** (navel of the Moon)	Tycho and ray system
5 **Carthusia** (Chartreuse)	Copernicus
6 **Hecates Penetralia** (Hecate's Shrine)	Mare Vaporum?
7 **Eoum Mare** (Eastern Sea; Eos=the dawn)	Oceanus Procellarum
8 **Boreum Mare** (Northern Sea)	Mare Frigoris
9 **Riphaeus**	Montes Apenninus
10 **Vallis Umbrosa** (Shady Valley)	?
11 **Rupes Nivea** (Snowy Cliff)	?
12 **Salinae** (Salt Pits)	?
13 **Amara Mons** (Bitter Mountain)	?
14 **Lacuna** (a pit or hollow)	?

[1] Originated with Plutarch, as noted earlier.

[2] **Homuncio** means 'little man', and clearly refers to the same group of maria described by Albertus Magnus, i.e. Mare Serenitatis and Mare Tranquillitatis, with Mare Nectaris and Mare Fecunditatis marking the legs.

    Thersite(s) – an officer; the most ugly, deformed and illiberal of the Greeks during the Trojan War. Killed by Achilles.

[6] Originated with Plutarch, as noted earlier. His description appears to refer to Mare Imbrium, as noted earlier.

[9] Conjectural mountains of northwestern Asia in mythology and old maps, on which they are usually spelled 'Rhiphaeus'.

REVIEW OF THE MELLAN ENGRAVINGS

One does not need to be a lunar expert to see that these images are infinitely more detailed, accurate, and aesthetically pleasing than any of their predecessors. For anyone who has looked at the Moon through binoculars or a small telescope, the usual remark on seeing these images is that they 'really look just like the Moon'. I would like to encourage any of my readers who have such optical aids to make the comparisons for themselves or, better yet, to try drawing what they see. Not easy!

Starting with the full-Moon image, the remarkably accurate rendering of the details and nuances of shading are quite surprising when one considers the difficulties faced by the artist. He was using a telescope made by Galileo, which gave an upright image but very restricted angular field of view – probably less than the lunar disc in diameter. There were no equatorially mounted and driven telescopes then, so that the lunar image would be continuously wandering out of the field of view, requiring frequent repositioning of the instrument. The full Moon rises at sunset, so that artificial illumination would be required – candles flickering in the breeze, maybe? Probably not, as we will see shortly.

A closer comparison between the full-Moon image and the frontispiece photo reveals a rather curious situation. The right-hand half of the image agrees very closely with the corresponding area in the photo; for the other half, however, the markings are all placed rather too high up on the disc. Comparing now the two quarter-phase engravings with that of the full phase, we find that the positioning of the albedo markings in the latter corresponds exactly with that in the former two. It seems likely therefore that Mellan did not draw the full Moon in candlelight – he just combined the relevant parts of the other two images, both of which could have been made in bright twilight, and omitted the shadows and shadings caused by the lunar topography. This tends to be confirmed by the note at the bottom of the image of the nine-day-old Moon: 'Cl. Mellan painted and engraved [this] phase [at] Aix [in the] year 1635 October 7 from daylight until the end of twilight.'

The apparent northward shift of the left half of the full-Moon image might well be due to the Moon's latitudinal libration, whereby the tilt of the lunar axis with respect to the Earth causes an apparent nodding motion of almost 7° each side of the mean position. Thus the image of the 22-day-old Moon was probably made when the Moon's north pole was tipped away from Earth, while for the nine-day image, it was near the mean position or tipped towards Earth. This is confirmed by comparing the phase images with photos taken at various librations in the north–south direction. Mellan's phase images were sufficiently accurate that he could not join the two together for the full Moon image without distorting the overlap area a little – a most remarkable testimony to the reliability of his draughtsmanship.

Five or six years after these three engravings were produced, Gassendi learned that Hevelius was commencing a similar but far more ambitious project in Danzig, and that P. Juan Caramuel y Lobkowitz (1606–82) was also contemplating something of the same sort. Gassendi wrote to Caramuel that he was happy to be relieved of the burden of trying to produce satisfactory names for the lunar features, and that the new selenographers would no doubt baptize them much more suitably. Caramuel replied that he proposed to name the 'promontories, islands and valleys' after contemporary learned persons. 'All our friends will be there,' he wrote, 'you yourself, and **Peiresc**, and **Mersenne**, and **Naudé**.' [my emphasis]. Further, in a letter to P. Anton Schyrle of Rhaetia dated 16 March 1643, Caramuel wrote that he would name a feature '**Promontorium Rheita**, to ensure Schyrle's immortality' [my emphasis]. Caramuel's nomenclature, as Gassendi's, never proceeded any further, due to the publication of three pioneering lunar maps with major nomenclature systems during the next eight years.

CHAPTER 3

VAN LANGREN (LANGRENUS) AND THE BIRTH OF SELENOGRAPHY

The impending activities of Hevelius and Caramuel posed a serious threat to yet another Moon-mapping project which, at that time (1643–4) had already been under way for about a decade. In 1628, the very same year that Gassendi saw the need for a lunar map and atlas, Michiel Van Langren (1600–75), member of a prominent Flemish globe- and map-making family, and 'Royal Mathematician and Cosmographer' firstly to Belgian royalty and later to King Félipe IV of Spain, came to exactly the same conclusion. The longitude problem again at first provided the chief driving force; he knew of the lunar eclipse timings method, but, seeking a more continuous celestial 'clock', conceived the idea of timing sunrises and sunsets on

> ...islands and the peaks of mountains most often isolated from the main continuum which in an instant appear on the face of the waxing Moon, and also those which suddenly vanish on the waning Moon, which first instant and duration is a help in finding longitude.

During a sojourn in Madrid in the early 1630s, Van Langren formulated plans to produce drawings of the lunar phases, a map, a globe with latitude and longitude circles marked and with provisions for setting the librations. He also planned to publish his theory of the librations, and instructions for obtaining terrestrial longitudes from actual sunrise or sunset timings on listed lunar features. At that time he had already decided upon the scheme of nomenclature that he would use, since it is referred to in a letter from King Félipe to Princess Isabella (Félipe's aunt),

37

dated 27 May 1633. Under his engraved map, which was published about April 1645, Van Langren repeats Félipe's words thus.

> It also pleased him [i.e. King Félipe] to have the names of illustrious men applied to the luminous and resplendent mountains and islands of the lunar globe, which might be used in the future in astronomical, geographical and hydrographical observations and corrections.

In the same letter, Félipe encourages Van Langren to proceed with his planned project, and authorizes Isabella, whose considerable interest in the subject provided the necessary impetus, to fund it 'within the moderation required by the present state of affairs.' (funding problems are apparently eternal and universal!). Unfortunately, Isabella died while Van Langren was returning from Spain in 1634; this not only severely affected his funding prospects, but also dampened his enthusiasm somewhat. However, unlike the Gassendi project which produced only three drawings of lunar phases before it collapsed due to the death of its sponsor, Van Langren did at least prepare 30 such drawings over the following nine-year period.

By 1643, Van Langren was fully aware of the possibility of being upstaged by Caramuel or Hevelius in the publication of some form of selenographical document. Thus, again under his map, he writes

> There was even the danger that since the work was being divulged abroad to a growing extent, some other person would appropriate it as his own and publicize it in his own name.

COMPOSING THE MAP

Goaded into action by this threat, he apparently wasted no further time in preparing a map from the 30 drawings of different lunar phases that he had already made. This was completed by early 1645, at which time he faced the problem of actually choosing the names to be used on the map, and of deciding how best to allocate those names to features of widely different prominence on the Moon's face. It seems that at this stage he had more features on his map in need of names than were contained in his list of 'illustrious men', and that he then decided to add the names of 13 saints in order to complete the allocations. On learning of this decision, his friend and mentor in Madrid, P. Jean-Charles della Faille, S.J., with whom he was in constant correspondence, asked Van Langren in a letter dated 28 January 1645 to do him and another mutual friend, Don Lorenzo Cocchi, the honour, if possible, of adding the names of two

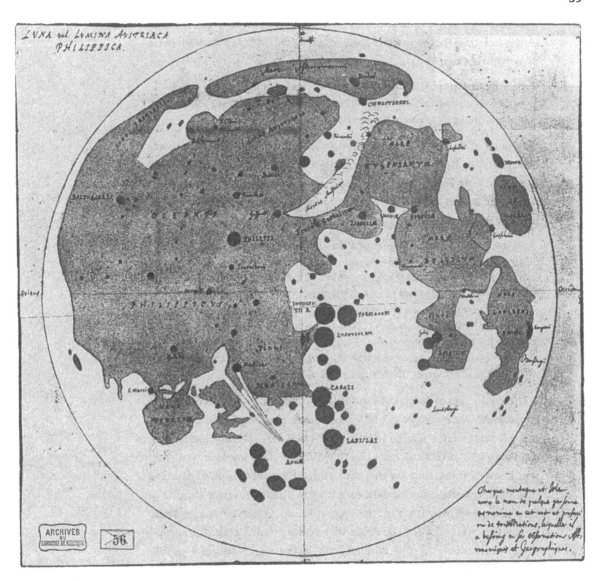

Fig. 25. Michiel Van Langren's manuscript map, a pre-publication document establishing his priority.

saints of their choice, namely **St Vincente** (Ferrer), and **St Demetrius**, respectively. The former name does indeed appear on the final map, together with the other 13 chosen by Van Langren himself, but the latter does not.

Before allocating the names to the formations that appeared on his engraved map, Van Langren prepared a simpler hand-drawn and coloured version and provided it with 48 names to indicate the general scheme of nomenclature that he would use (fig. 25). Of these names, 47 were used again on the engraved map, but many were moved to different features (Appendix A). In the lower right-hand corner of this MS map we read, in his own hand (in French)

> Each mountain and island will have the name of some person renowned in this art and profession of all nations; which is necessary for their astronomical and geographical observations.

The name of **St Vincente** appears on this MS map, showing that it was drawn after he had received the suggestion from della Faille. On the other hand, the long legend below the engraved map must have been almost completed by this time, since no mention is made of the inclusion of the names of saints. There, he says that he has used

> ... the proper names of Kings and Princes (who reign today in Europe and are patrons, protectors and promoters of the Sciences and Mathematics), and with the names of other people old and recent who excel in this knowledge and call down praise and fame on themselves, as sterling monuments to their wisdom; wherefore we will also publish a book in their honor. We are very sorry that as yet we have not been able to transcribe (although we hope to in the very near future) the names and merits of others who are otherwise famous in these arts, so that we might equally inscribe them in our resplendent globe.

The final stages of putting the names and other finishing touches on the copperplate were completed by early March 1645, and printed copies made from it were available shortly afterwards (fig. 26). The Moon image has a diameter of about 34 cm, as on the MS map. There are at present only four known extant copies of this map, suggesting a very limited printing run, although others may well lie forgotten in private libraries. It exists in three slightly different states; the copy at Leiden Observatory appears to be the first state, that of the Royal Observatory, Edinburgh, the second state, while those of the Bibliothèque Nationale, Paris, and of the Observatorio de Marina, San Fernando, are the third state. The small differences between these are given in Appendix B.

THE STRASBOURG FORGERY

There also exists another engraved map, obviously copied directly from a third-state example of the original, in the Library of the University of Strasbourg. This copy, which is devoid of the five quotations and lengthy explanatory text of the original, is clearly a counterfeit map, not only because of the missing legend, but also because the engraving is inferior and the transcription sloppy, several names being mis-spelled or absent.

Fig. 26. (opposite) The first real map of the Moon, by Van Langren, published in 1645. It depicts both the surface shadings and topography, and introduces the scheme of names used today. Some 168 of his 325 names have survived but, except for four of these, were soon moved to different features.

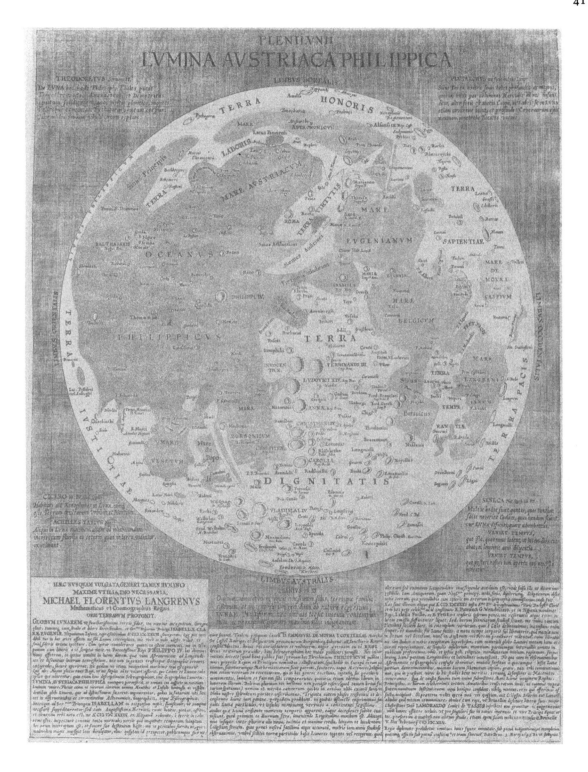

An intriguing point is that six of Van Langren's original names have been replaced by entirely new names – see fig. 27 and Appendix C.

Van Langren certainly foresaw the possibility of forged copies and name changes, since under his map we read

> But lest confusion arise in Astronomical and Geographical observations by someone perhaps changing the names of these areas on the Moon, we have freely communicated to the whole world the great bounty of these plans, by which we now venerate those men famous for these studies and their defenders and promoters. For this grave reason, with total submission we produce this image of the Moon, dedicated to the Kings, Princes, and most famous lovers of these arts, and we beg that just and good persons preserve the proffered list of names and change nothing, and that they receive it in the spirit that we present it.

Having appealed to the better sides of people's natures to leave the names intact to prevent future confusion, he finally adds

> By Royal Decree changes in the names of this map are forbidden under pain of indignation, and copies and other forgeries are forbidden under pain of confiscation and a fine of three florins.

This appeal, together with the prospect of the 'pain of indignation . . . confiscation and a fine of three florins' apparently failed to impress whomever it was who made the Strasbourg forgery. It also failed to prevent the appearance, within a span of only six years, of two more schemes of nomenclature, those of Hevelius (1647) and Riccioli (1651). Thus was the subject of lunar nomenclature doomed to confusion right from the very outset.

REVIEW OF THE VAN LANGREN MAP

A careful study of the final map shows that it is more accurate than might be judged from the monotonous shading (stipple) and somewhat stylized outlines of the darker areas. Craters are shaded as if illuminated by a morning Sun, a technique still used today on the best lunar maps. Trying to identify all the features on the map (Appendix D) is an interesting but rather tedious exercise, requiring a fairly extensive set of whole-disc lunar photographs taken at many different phases for making comparisons. This leads to the conclusion that isolated peaks or craters, and small light or dark spots are often not differentiated from each other. It is easy to understand Van Langren's reasoning here: isolated larger peaks in the mare areas show as bright spots at or near full Moon, as do the inner walls or, in many cases, the entire interiors of numerous craters. A few craters

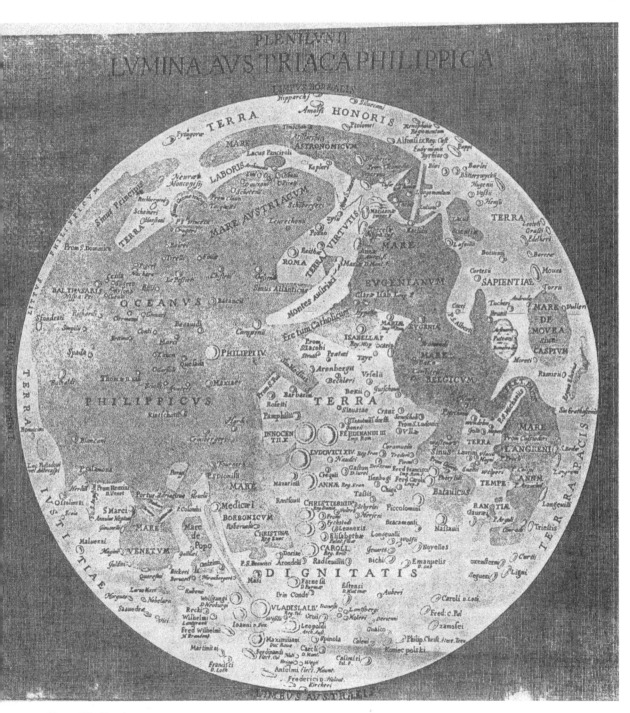

Fig. 27. The Strasbourg
forgery of Van Langren's
map.

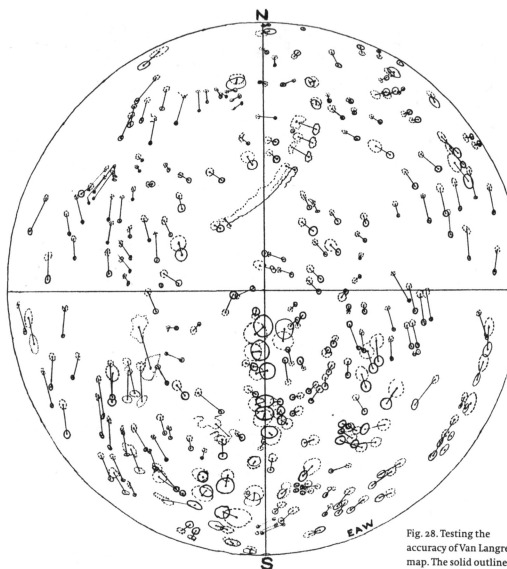

Fig. 28. Testing the accuracy of Van Langren's map. The solid outlines are traced from the map while the correct sizes and positions are dashed. It can be seen that positioning errors in latitude are much larger than those in longitude, suggesting that he used the moving sunrise and sunset lines as effective measurers of longitude.

have dark floors. So might it not be reasonable to assume that some of the smaller light or dark spots are actually hills or craters that are too small to be detected as topographic features?

About 13 features could not confidently be identified even when allowances were made for reasonable errors of placement. Figure 28 illustrates a by-product of this exercise, a tracing of the map overlaid with the same features from a modern mean-libration map. It can be seen that placement errors are mainly in lunar latitude, suggesting that Van Langren used the regular advance of the terminator to obtain fairly accurate longitude positions. The errors in latitude may well have resulted from ignoring the libration in latitude, as apparently was the case for the Mellan images.

VAN LANGREN'S NOMENCLATURE
Van Langren indeed used the names of European royalty and nobility, philosophers, scientists, mathematicians, patrons etc., but we also find the names of explorers, religious leaders, the 14 saints, and possibly other categories if all the names could be identified. In addition, most of the 'watery' features have geographical or descriptive names such as **Mare Venetum** (the Venetian Sea), **Sinus Geometricus** (Geometric Bay), **Portus Gallicus** (French Harbor); also used are the general terms **Oceanus, Lacus, Fretum, Flumen, Fluvius**, and **Aestuaria** (Ocean, Lake, Strait, River, River, and Estuary).

The names of the main divisions of the highland areas, all given the general term **Terra**, reflect desirable human qualities such as dignity, honour, justice, wisdom etc. There is also one **Littus** (Shore). Van Langren also uses the terms **Promontorium** (Cape) and **Montes** (Mountain Range), and one **Annulus** (Ring). Craters, isolated peaks, and small light and dark spots are simply assigned names without any qualifying word such as **Mons**, but the genitive case is always used with the latinized version of a name; thus **Hugenii** means the Mountain or Island of Huygens, judging by the MS map note.

At first sight, the proper names appear to be distributed more or less at random over the Moon's face, except that the larger and more prominent formations have received the names of kings, queens, members of royal families, the Pope, a cardinal and so on. However, a closer look shows that most of the astronomers are clustered in or near **Mare Astronomicum**, and moreover, that the older Greek and Arabic names (e.g. **Ptolomei, Albategni**) are situated nearer to the northern limb than the more recent and contemporary astronomers. This latter arrangement was adopted and extended by Riccioli in 1651, but without any acknowledgment to Van Langren for first conceiving the idea. It is interesting to note that Van Langren assigned his own name to a prominent crater (**Langrenus**), and also to the adjoining sea (**Mare Langrenianum**); the former assignment stands to this day, but the latter was quickly superseded, as we will see. Other assignments that have remained unchanged from this map are

Endymionis and **Pythagorae**, apart from the dropping of the genitive case. We see that **Sinus Medius** (Central Bay) appears as Sinus Medii (Bay of the Center) on our modern maps, but this appears to be more likely the re-introduction of an obvious name, as we will see later. As noted earlier, the map shows our Mare Crisium as **Mare Caspium** (as an alternate name to **Mare de Moura**), confirming that this older appellation was still extant at this time. It also shows our Grimaldi as **Lacus Possidoni** or **Antecaspi** [*sic*], whereas our Mare Humorum would seem to be the more logical identification, as assumed for the Gassendi nomenclature.

Two typefaces are used for the names on the map – Roman and Italic. The former is used in capital letters for the highlands divisions, the chief seas, and for the kings, queens, emperors and the Pope. Roman lowercase type seems to be reserved for other members of royal families and the nobility, and the saints; italicized lettering appears to indicate scientists, mathematicians, cartographers and others in those general disciplines. Of the 325 names on the map, some 68 are still in use on our modern maps but, apart from the four cases noted above, in different locations. These, plus 7 more that were used by later mappers but not today, are noted by asterisks in Appendix D, which lists the complete nomenclature with the modern names for the features, where such are in existence. Note that **Reithae=Schyrlei** (i.e. same person), and **Brahei** appears as **Tycho** nowadays.

CHAPTER 4

SIX MORE YEARS OF BUSY ACTIVITY

RHEITA AND FONTANA

Shortly after the publication of the Van Langren map, the Capuchin friar and optician, Anton Schyrle of Rheita (Rhaetia), (1597–1660), included a lunar map in his *Oculus Enoch et Eliae, sive Radius Sidereo-Mysticus* [*The Eye of Enoch and Elias, or the Mystic Star-Ray*] (fig. 29). Note that this image is presented south-up, a result of his using a positive eyepiece rather than the negative (concave) lens used by all of his predecessors except Fontana. Letters placed on some of the surface features are references for the accompanying text, as noted earlier.

The map is a copperplate edition, diameter about 18.5 cm, of an original drawing made by Rheita. It is essentially a full-Moon image with a very few craters included from observations made at other phases. The ray systems emanating from Tycho and other craters are very stylistically drawn, and bear little resemblance to reality. Indeed, those lettered H and V are non-existent. The small bright spots are almost entirely randomly placed.

The work of Francesco Fontana (1585–1656), whose early drawings and 1646 book were referred to in chapter 2, includes 24 sequential images of recently observed (5 October 1645 to 1 January 1646) lunar phases, plus the two earlier ones reproduced in fig. 22, and two others dated 1630 and 1640. These images, with diameters from 10 to 14 cm, and produced on copperplates by the author himself, are of low accuracy and add virtually nothing to lunar science of that time. His descriptions of some of the lunar features are quaintly interesting; thus he refers to 'margarituncu-lae' (little pearls), 'gemmales' (gems), 'fonticulus' (little fount), 'pila nigra' (black hairs), and uses other poetic names. Figure 30 illustrates two of his

Fig. 29. Rheita's Moon map. although it is basically a representation of full phase, south-up, a few topographical features are shown. The rays from Tycho and other craters are highly stylistic.

a

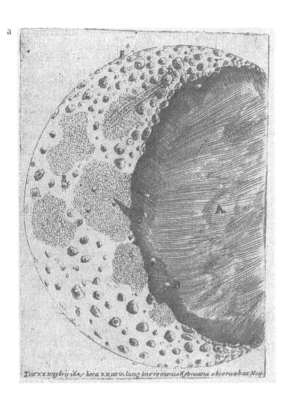

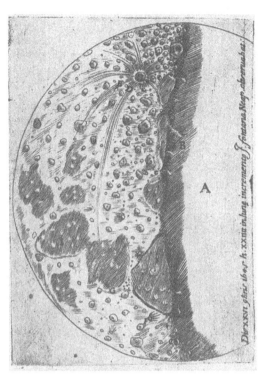

Fig. 30. Two of Fontana's wildly imaginative (and inaccurate!) images of the waxing Moon (a), and his image of the full phase (b). Note the pairs of parallel rays emanating from Tycho and Kepler craters, reminiscent of the nonexistent but confidently drawn twin 'canals' on several eighteenth century Mars maps.

b

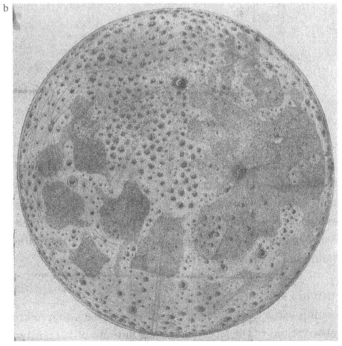

SELENOSCOPIA

OVERO

Aftronomicofifica fpecolatione circa la Luna,

DOVE

Curiofamente fi efaminano del notturno gran Luminare
le conditioni più effentiali, e le accidentalità.

Colle ponderationi Aftrologiche delle varietà de'tempi,
ed altre confeguenze, che poffono fuccedere
nell'Anno 1647.

Aggiuntoui vn'Indice diurno delle congionture buone, ò ree per le
operationi più importanti in materia medicinale princi-
palmente, ed Economica d'ogni forte.

Di Ouidio Montalbani il Rugiadofo Academico della Notte,
e frà gl'Indomiti lo Stellato.

All'Altezza Serenifs. del Gran Duca di Tofcana

FERDINANDO SECONDO

In Bologna, preffo Gio. Batt. Ferroni. Con licenza de' Superiori.

Fig. 31. Title page of a small book that uses one of Fontana's early images with a few of Van Langren's names added, except that 'Terra Fortitud.' is not on any of his maps!

images of the crescent Moon and another image of full Moon (diameter 24 cm); they are south-up, once again resulting from the use of a positive eyepiece. Figure 31, the frontispiece of a small book by Ovidio Montalbani (1647), shows a poor copy of Fontana's 1629 image, but with a small selection of Van Langren's names added. Note in particular **Terra Fortitudinis**, which is certainly not in the Van Langren maps.

HEVELIUS AND HIS *SELENOGRAPHIA*

Johannes Hevelius (Höwelcke), who lived from 1611 to 1687, commenced systematic observations of the Moon in the late autumn of 1643. After a period of about a year and a half, he had made sufficient drawings of lunar phases to be able to combine the portrayed features into whole-disc maps. Unlike the less fortunate Van Langren, he had substantial financial resources available from the family business (brewing), and was able to combine all his observations into a bulky tome titled *Selenographia*, which was published in 1647 in sufficient quantity to be widely available. This contains three major maps, each being about 29 cm in diameter; one is an image of the full Moon, depicting the surface markings as seen to best advantage at that phase ('P', fig. 32); a second shows the same markings, but with the addition of the topographical features as seen at the various other phases ('R', fig. 33); the third map is for the nomenclature, and depicts the topography in the 'rows of termite hills' convention used on terrestrial maps of the time. Hevelius himself engraved the first two of these maps, but the last was done by a professional engraver (Jeremiah Falck), who also did the corner cartouches in all three ('Q', fig. 34). Hevelius also includes engravings of 40 different phases, arranged in order of increasing lunar age, also prepared by himself (fig. 35), with descriptions of those formations that were of particular interest, along with observations of eclipses and of the librations of the Moon, attempts to measure the Moon's axial tilt, plus many observations of sunspots, the planets, Jupiter's satellites etc.

HEVELIUS'S NOMENCLATURE

Before dealing with this subject, which he does at considerable length, Hevelius lists many reasons for his belief that the Moon's brighter regions are actual land and the darker ones water, capping them off by quoting Kepler's revised pronouncement to the same effect. He introduces the subject of lunar nomenclature by presenting many arguments for naming the lunar features, and gives as precedents the naming of the planets, stars, constellations, and geographical features. He writes

> What, I wonder, would the science of astronomy be like, if we could not properly discriminate among the stars themselves. Without the use of unique names, all observations, both ancient and modern, would be useful to nobody, and the books describing these things would seem to us to be more like enigmas rather than descriptions and explanations.

Further on he writes

> ... lastly I confess I have taken on an arduous task, **hitherto unheard of**, of applying certain names to parts of a remote heavenly body, which has not been done by anyone to this day, as far as I know, a work that is tentative and happily not yet absolute (my emphasis)

Fig. 32. Image of the full Moon by Johannes Hevelius (1647), an excellent effort that compares favorably with reality – see the frontispiece. The double circle represents the limits of the lunar librations that occurred during his observations.

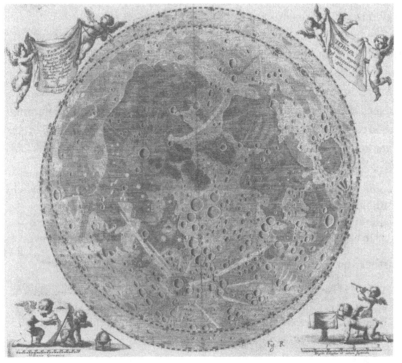

Fig. 33. Map by Hevelius that combines full-Moon markings with observed topographical details. Many of the latter, as well as not being aesthetically pleasing, are often quite inaccurate.

Fig. 34. Hevelius's map for his nomenclature. The surface is depicted as a disc sloping away from the viewer, and hills, mountains and crater rims are shown as rows of 'termite hills', following the then current geographical practice. Note that several crater ray systems are inexplicably represented as mountain chains.

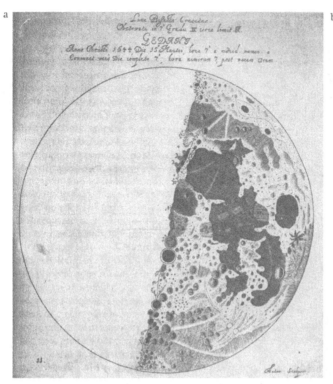

The words that I have emphasized do not agree with what Van Langren wrote in a letter to Ismael Boulliaud, dated 'the day of the summer solstice 1652'

> ... who [i.e. Hevelius], in my opinion, was wrong not to mention my work which he saw fully two years before he published his selenographic production.

Fig. 35. One of Hevelius's 40 engravings of different phases, this one (a) of first quarter. A photo (b) of a closely similar phase is included for comparison purposes.

It is possible, of course, that Van Langren's map never reached its destination. In any case, Hevelius says that at first he thought of applying the names either of men of the past of surpassing virtue, or else of the names of recent and contemporary famous and most learned men in mathematics. As examples, he gives the following.

Oceanus Coperniceus	Mons Mersenni
Lacus Galilaei	Fretum Eichstadianum
Peninsula Gassendi	Mare Kepplerianum
Sinus Wendelini	Insula Scheineriana
Promontorium Crugerianum	Vallis Bullialdi
Oceanus Tychonicus	Desertum Linnemanni
Palus Maestlini	

He then continues at some length to explain why he decided against this scheme; put succinctly, he feared that he would invoke jealousy and enmity from contemporary scientists who might feel slighted by having their names assigned to inferior formations, or by being omitted entirely.

Looking around for a 'safer' and 'more fitting' scheme, he finally decided to use classical names of terrestrial seas, islands, mountain ranges, countries etc. He continues

> So I brought myself round to this prospect, if there are such Geographical names, and they are well known, to apply them to the places on the Moon provided that (which must first be investigated) the hemisphere of the Moon facing us could be fittingly ordered to a certain part of the Earth's globe, in regard to the siting of the locations.
>
> Immediately I had put my mind to this work of ferreting out an answer, and had contemplated practically all of Geography, I found to my perfect delight that a certain part of the terrestrial globe and the places indicated therein are very comparable with the visible face of the Moon and its regions, and therefore names could be transferred from here to there with no trouble and most conveniently; namely, think of the part of Europe, Asia and Africa that surround the Mediterranean Sea, Black Sea and Caspian Sea, and all the other regions including and adjacent to them, which are: Italy, Greece, Natolia, Palestine, Persia, a part of Sarmatia and Tartary, Egypt, Mauretania etc., which places extend from the tenth to the ninetieth degree of longitude and from the twenty-fifth to the sixtieth degree of latitude.

Hevelius proceeds to congratulate himself on hitting upon such a perfectly apt scheme, explaining his choice of the old classical names of geographical features rather than the current ones, saying that he '. . . could never get over wondering how all this turned out so well, as I never thought it would in the beginning.' He later suggest that '. . . those studying Selenography might learn all these applied names with great pleasure and little labor and so commit them to memory; . . .'. As it later turned out, it was the length, awkwardness, and archaic appearance of these names, amongst other reasons, that eventually ensured the demise of the scheme. Thus it is a little difficult to 'commit to memory' such names as **Montes Coibacarani, Celenorum Tumulus, Chersonnesus Taurica, Lacus Corocondametis, Lacus Hyperboreus Inferior,** and many others of similar length.

He used many general terms for this nomenclature, as listed here.

Caput	headland, cape	**Catena**	chain
Chersonnesus	peninsula	**Collis**	hill

Continens	continent	**Desertum**	desert
Eruptio	outbreak	**Fluvius**	river
Fons, Fontes	source, sources	**Fretum**	strait
Insula	island	**Lacus**	lake
Mare	sea	**Mons**	mount
Montana	mountainous area	**Montes**	mountain range
Palus	marsh	**Petra**	rock
Planitia	plain	**Promontorium**	promontory
Regio	region	**Scopulus**	sea crag
Sinus	bay	**Stagnum**	swamp
Tumulus	mound	**Vallis**	valley

Such a profusion of different terms did not help with the viability of the scheme. Other characteristics that detracted from it were Hevelius's tendency to apply a single name to a group of several craters – a distinct drawback when a single crater in a group is being described – and an inexplicable aberration where he named several bright surface streaks as though they were mountain ranges.

The nomenclature map is accompanied by a list of named features and areas, but the list and map do not quite agree with each other. The complete listing is given in Appendix E, with 286 entries. Of these, only ten appear on modern maps, with six having been moved to different formations by later selenographers. These ten are listed in Appendix F.

REVIEW OF THE HEVELIUS IMAGES AND MAPS

By comparing the phase images (e.g. fig. 35) (which were engraved from observations made from November 1643 to April 1645), with photos at similar phases, we find that the topographical features brought into relief at and near the terminator, i.e. the sunrise or sunset line, are reasonably accurate. Features further from the terminator are less trustworthy in their portrayal or even existence.

The full-Moon image (fig. 32) is a quite remarkable representation, as can be seen by comparing it with the photo (frontispiece). Both the proportions of the features and their relative shadings are reasonably close to reality, although the overall appearance is less aesthetically pleasing than the Mellan image (fig. 24). This is undoubtedly another reason why Van Langren's map, with its highly simplified mare outlines and sparse indication of albedo markings, could not compete against Hevelius's production.

Turning now to the map (fig. 33), we find the situation rather less convincing. The percentage of non-existent craters is greater than in Van Langren's map, and a number of the objects depicted as craters are, in fact, peaks or ridges. Hevelius used Van Langren's convention for crater portrayal, but with reversed (sunset) illumination. The hard crater outlines are rather less pleasing in appearance than the softer ones used by Van Langren. The map for the nomenclature, fig. 34, is based on the previous map except that, as already mentioned above, Hevelius shows the bright rays as mountain ranges.

All three maps will be seen to have a double circular edge, or limb. Hevelius was fully aware not only of the Moon's libration in latitude, but also that in longitude, a phenomenon resulting from the Moon's constant rate of rotation but variable rate of revolution around the Earth. The double limb was an effective way of depicting the areas of the north and south limbs that were alternately brought into view and moved out of view by the libration in latitude. Note that the displacement is not exactly in the north–south direction, but includes some east–west libration as well, which Hevelius certainly observed, and recorded in this manner.

The librations were first observed by Galileo, that in longitude in 1632, and in latitude five years later, but Hevelius was the first to portray the effects of these in images. Actually, Harriot was the first to observe the libration in latitude. Observing the Moon on 14 December 1611, he writes 'I noted that the darke partes of 28, 26 were nerer the edge then is described', i.e. in his full Moon map – see fig. 14. He made no further remarks, probably thinking that he had merely placed that dark marking, now named Mare Frigoris, too far from the limb of the Moon on his map.

DIVINI AND SERSALE
Eustachio Divini (1610–85), a telescope maker from the Naples area, published a broadside in 1649 containing a large (diameter 28 cm) image of the full Moon, a small one of the crescent Moon, and images of Venus, Jupiter and satellites, and Saturn, as observed from Rome in that and earlier years (fig. 36). In the text at the top, he says that the full-Moon image was made using telescopes of 24 palms – about 17 feet, and 16 palms – about 11.5 feet – the focus of the latter being provided with a reticle of fine threads to act as an aid in positioning the details correctly with respect to each other. The larger instrument was used to scrutinize the Moon for the 'smallest and most minute spots'. However, this image, supposedly made in March of 1649, repeats so many idiosyncrasies of Hevelius's image (fig. 32) that Divini obviously started out with that as a

Fig. 36. (opposite) Moon map by Eustachio Divini, 1649. Although he claims to have prepared it from his own observations using an eyepiece reticle, this is clearly based on Hevelius's full-Moon image (fig. 32) with a few cosmetic changes and meaningless additions. Two years later, Riccioli named a crater for him.

Fig. 37. Image of the full Moon by Hieronymus Sirsalis, 1651. Although mentioned in the heading for Grimaldi's map (fig. 38), this map lay unknown until listed in a recent library catalogue.

base and simply made additions and amendments as he noticed omissions or differences from what he observed. The text also notes that he did not use a concave lens, with which an enlarged image of Moon and reticle would not have been possible. His positive eyepiece or lens would have inverted the image, but he has presented it north-up.

Gerolamo Sersale (Sirsalis, 1584–1654), a Neapolitan Jesuit, also produced an image of the full Moon (diameter 34 cm) from an observation made in July 1650; this was published as a broadside early in 1651 (fig. 37). While the general proportions of the larger features are seen to be fairly good, the smaller details such as the numerous 'bays' in the edges of the maria, the isolated small dark areas, and the small bright spots and rings

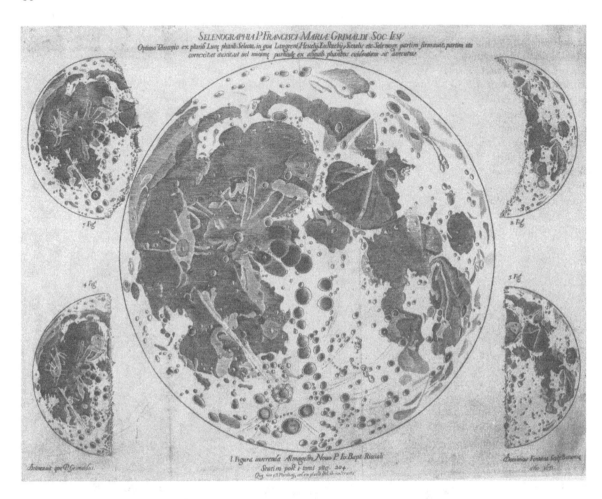

are surprisingly well portrayed. Despite this, the image does not have the aesthetic appeal of the Mellan full-Moon engraving. Neither this nor the Divini map supplied any nomenclature.

RICCIOLI AND FURTHER CHANGES

Hevelius's *Selenographia* was rightly hailed as an outstanding contribution to the study of the Moon, with its particular emphasis on the formations of the lunar surface as illustrated by the aesthetically pleasing maps and images of the different phases. Its success may be judged from the fact that it remained the foremost source of information on the subject for almost 150 years. The nomenclature, however, enjoyed no more than four years of precedence when yet another scheme made its appearance. In

Fig. 38. Francesco Grimaldi's 1651 Moon map, based on the earlier maps of Van Langren, Hevelius, Divini, Sirsalis, and others, but 'corrected and augmented' by his own observations, made with 'the best telescope from many phases'.

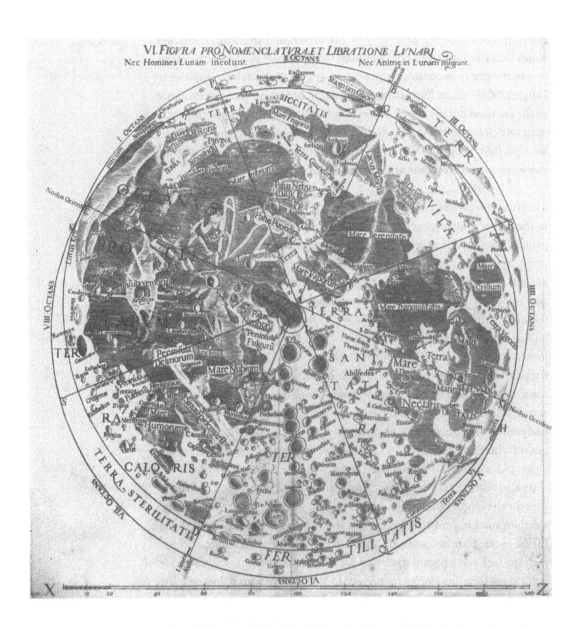

Fig. 39. Riccioli's nomenclature on Grimaldi's map. The majority of names and their allocated formations remain unchanged to this day, and many of the unnamed craters are readily identifiable.

1651, P. Giovanni Riccioli, S.J. (1598–1671) published a large, two-volume work on astronomy titled *Almagestum Novum*, that included a section on the Moon with two new maps – one showing the surface markings and topography unhampered by names, the other being similar but with his new nomenclature added. These are illustrated in figs. 38 and 39, respectively; the originals have a diameter of 28 cm. In the former, the heading notes that the map is the work of P. Francisco Grimaldi, S.J. (1618–63) (who is better known for his discovery of the diffraction of light). It further states that the 'best telescope' was used to view many phases of the Moon as a basis 'partly to confirm, correct and augment the selenographies of Langrenus, Hevelius, Eustachius, Sirsalis and others, so that the evidence of the smallest details at any phase may be followed up.' Above the latter map, which notes that it is '. . . for the lunar nomenclature and libration', we read that 'People do not inhabit the Moon, neither do souls migrate there', refutations of some ancient beliefs recorded by Plutarch.

RICCIOLI'S NOMENCLATURE

In the text accompanying the maps, Riccioli acknowledges Van Langren's priority over Hevelius; he notes that there are differences between the maps, specifically in the schemes of nomenclature adopted by those authors. He writes rather slightingly of Hevelius's scheme because of the very poor correlation between the markings of the Moon's face and the features of that part of the Earth chosen by Hevelius. Instead, he says that he has adopted the general scheme of Van Langren, but rather than including the names of current ecclesiastical and other dignitaries, or friends, he has restricted his list to persons ancient and modern who have contributed to, or have had some intimate connection with astronomy. He actually used 63 of Van Langren's names, all but three (**Pythagoras**, **Endymion**, and **Langrenus**, as noted earlier) being located on different lunar formations. He added 147 new names which were selected from the biographical list in his *Almagestum Novum*.

He also notes that rather than using Van Langren's choice of moral qualities for the larger lunar areas, he has chosen names related to terrestrial weather, since it was believed that the Moon, being the closest body to the Earth, exerted an influence on that element. While Van Langren could perhaps seek some consolation that Riccioli had chosen his basic scheme over that of Hevelius and had commented favourably about it, he was quite understandably irked that his pioneering nomenclature, which he might reasonably have expected to remain the basis for later improvements and additions, had been twice superseded in only six years. Thus, in the same letter to Gassendi in which he complained of Hevelius's total disregard of his map, Van Langren adds, 'Again, here is Father Riccioli, professor of Bologna, who has changed everything even though he had nothing but praise when I sent him my selenography.'

DISTRIBUTING THE NAMES

On the subject of the distribution of the names over the face of the Moon, Riccioli writes that these are not placed randomly as in Van Langren's map, but are mostly grouped by various criteria. We have already seen that several of Van Langren's names are collected into logical groups, so that Riccioli erred in this statement. In any case, he describes his placement scheme in some detail. It will be seen from his map (fig. 39) that the map is divided into eight parts (octants), and many of the names have been placed with respect to the boundaries of these. Thus, towards the top of the map he says that he has placed the ancient astronomers, with most of the physical astronomers being in octants 1 and 2, and the rest of the ancients in octants 3 and 4. The more recent astronomers are placed in the lower part of the Moon, in octants 5, 6, 7, and 8. Furthermore, he notes that he has grouped the names by affinity, similarity in epoch, studies, philosophy, and so on. Thus he places **Meton** along with **Euctemon**, his observing colleague; **Plato** with his associates **Timaeus**, **Theaetetus** and **Archytas**, and not too far from **Aristoteles**. Similarly, he writes, **Eudoxus** is near **Calippus**, **Atlas** near **Mercurius**, **Julius Caesar** near **Sosigenes**, **Hipparchus** near **Ptolemaeus**, with **Albategnius**, **Alphonsus**, **Arzachel** and **Thebit** nearby. We note that these are all grouped near the central meridian of the Moon, as are **Regiomontanus** with his tutor **Purbachius** and pupil **Walterus**. Equally, **Tycho** is near **Gulielmus Hassiae**, **Hainzelius**, **Sasserides** and **Longomontanus**. Finally, he says that he has grouped **Copernicus** with **Rhaeticus**, **Moestlinus**, **Reinholdus** and the many others of their philosophy (i.e. a stationary Sun and an orbiting Earth) and thrown them into the stormy ocean like floating islands.

This last statement raises an interesting speculation. Riccioli was a member of the Society of Jesuits, and was thus obliged to adhere to the Church's teaching that the Earth stood still while the Sun moved, an apparent conclusion stemming from the literal interpretation of certain biblical texts. The complicated frontispiece of Vol. 1 of Riccioli's large work shows the Muse Urania weighing the Copernican and Tychonic World systems in a balance (fig. 40), with the latter overbalancing the former. Ptolemy, with his earlier system of Sun and planets orbiting a central Earth, are cast aside on the ground.

Riccioli goes to great lengths to list reasons why the Earth cannot be moving, and why the Tychonic system is the correct one. In this system, which is a hybrid between the Ptolemaic and Copernican systems, the Earth stands motionless at the centre of everything, with the Moon and

Fig. 40. The frontispiece of Riccioli's *Almagestum Novum*, showing the abandoned Ptolemaic Solar System with the Earth at the centre lying on the ground, while the hybrid Tychonic system outweighs Copernicus's system with the Sun at the centre.

Sun moving around it, while the other planets orbit the Sun. Thus, in looking at the crater **Tycho** and its bright ray system at the full phase (see the Frontispiece), we are being perpetually reminded that the light of true knowledge emanates from that astronomer and shines into the minds of all the other astronomers!

In the Ocean of Storms (**Oceanus Procellarum**) we do indeed find **Copernicus, Keplerus, Rhaeticus, Reinholdus, Lansbergius, Galilaeus, Seleucus, Cusanus** and **Aristarchus** – all of whom believed that the Earth and planets revolve around a central Sun. But we note that the crater **Copernicus** is larger than **Tycho**, and even though its ray system is less extensive than Tycho's, it is at least as prominent because of its isolated position amid dark maria. **Kepler** crater is noticeably smaller than **Copernicus**, but is almost as prominent under a high Sun because of its large surrounding nimbus (Riccioli's **Insula Ventorum**). Furthermore, the astronomer Kepler used Tycho's observations of planetary positions to formulate his three laws, and we note that his crater is situated on the continuation of one of the major Tycho rays. The **Galilaeus** of Riccioli's map is not the miserable little crater on modern maps, but the conspicuous bright spot now labeled Reiner Gamma (mistakenly drawn as a crater by Grimaldi). Lastly, we see that **Aristarchus**, farther north and thus nearer to his contemporaries, is easily the most brilliant object in the Ocean, with a ray system not much inferior to those of **Copernicus** and **Kepler**, and with both of which it intermingles – thus linking all three names for eternity.

I believe that there is considerably more behind these carefully chosen designations than Riccioli passes on to us in his allusion to 'floating islands'. His whole nomenclature is executed so carefully that the prominence of these three craters and their surroundings is not at all in keeping with his professed rejection of the Copernican system. I am personally convinced that, although he dare not admit it, Riccioli inwardly believed that the Copernican system was the correct one. By placing these three names on such eye-catching objects, I believe that he was quietly passing on this fact to future generations, and at the same time ensuring that these three astronomers would enjoy their rightful prominent place among all the other names on the map. And as Aristarchus was the first to postulate the revolutionary idea of the Earth orbiting a central Sun, to him naturally went the brightest crater.

Riccioli uses fewer general terms on his map than either Hevelius or Van Langren. Thus in the 'watery' category he includes only **Oceanus, Mare,**

Sinus, **Palus**, **Lacus**, and **Stagnum**, all previously used either by Van Langren or Hevelius. For the 'land' areas he uses only **Insula**, **Terra**, **Littus**, and **Peninsula**; conspicuously absent are the terms Mons and Montes. The lighter and darker lunar areas are seen to be named not only after various states of the climate and weather, but also after their effects on the land (terrestrial) and humanity. In general, the gentler and more benign areas are on the right half of the disc, while the left half has the more inclement aspects.

Riccioli provides a list of the 243 names used on the map, divided into the octants. However, the map has 247 names, and one of the listed names (**Arzet**) is not on the map! Furthermore, the name **Moestlinus**, which Riccioli says he has placed in the Ocean of Storms, is not in either the map or the list. Similar shortcomings have plagued lunar nomenclature ever since, with things coming to a head at the beginning of the twentieth century, as we will see later. We will also see later how Riccioli's nomenclature formed the core of the names found on our modern maps. The full list of his names is given in Appendix G, with changes in spelling, changes in identified feature, and various other correlations.

REVIEW OF THE GRIMALDI/RICCIOLI MAPS

It is interesting to compare the Grimaldi map (fig. 38) with its predecessors and the photo in the frontispiece. Under the map we read that it '... is not the full Moon, but is constructed from many phases'; thus it is a complete lunar map with both surface shadings and topographical features depicted. The former are quite detailed and reasonably well proportioned, but the heavier engraving style is less pleasing than that in Hevelius's image (fig. 33). Craters on the left half of the map are shaded as though illuminated from the left (east in the sky), while those on the right have the opposite illumination. The only mountains drawn as such – once again as a series of 'termite hills', are those on the right edge of **Mare Imbrium**. Van Langren named these '**Montes Austriaci**' (Austrian Mountains), Hevelius called them '**Mons Apenninus**' (Apennine Mount), and Riccioli '**Terra Nivium**' (Land of Snows). As noted in Appendix F, this is one of only four cases in which Hevelius names have prevailed to this day on their allotted formations.

Careful analysis of the topographical features shows that they are generally more reliably portrayed and placed than in either Van Langren's or Hevelius's maps, and many more small craters are shown. Thus here we see almost two dozen craters with central peaks or with small craters on

Fig. 41. Comparing the three pioneering maps with reality. A portion of Van Langren's map is shown in (a); (b) is the same area as depicted by Hevelius, and (c) shows Grimaldi's version, while (d) is from a modern photomap.

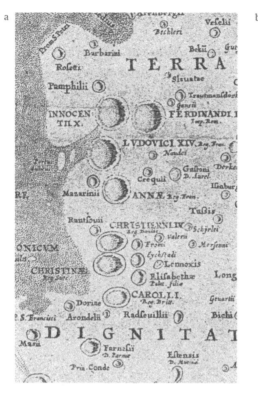

a

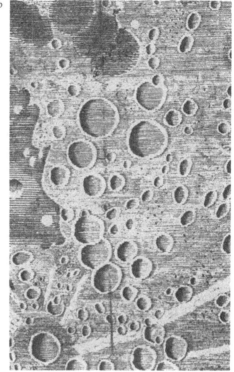

b

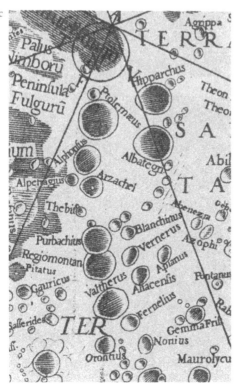

c

d

their floors; Hevelius shows only one, as does Van Langren. The one area that is less accurately portrayed than in the other two maps is that lying north of Mare Crisium, where confident identifications are difficult. Figure 41 shows an area just below the centre of the Moon's disc as portrayed in (a) Van Langren's map, (b) Hevelius's ('R') map, and (c) Grimaldi's map, with (d) a modern photomap for comparison. See fig. 65 for portrayals of the same area by Cassini (1679), Mayer (1749), and Russell (1805).

SECOND ERA

FROM ARCHETYPE TO MATURITY

CHAPTER 5

....

140 YEARS OF SPORADIC ACTIVITY

TWO COMPETING MAPS AND NOMENCLATURES

The wide distribution of Hevelius's *Selenographia* and Riccioli's *Almagestum Novum* in Europe completely eclipsed the very limited production of Van Langren's broadsheet map, and undoubtedly provided many people with their first authoritative descriptions of the Moon's surface features, complete with detailed maps and comprehensive nomenclatures. Obviously, readers must have compared the cartographic efforts of these two authors at that time, and could not have failed to notice the many differences between them. Thus the artistic maps and images of Hevelius contrasted sharply with the rather harshly portrayed engravings in the Riccioli work – yet the simple and logical scheme of formation names in the latter was clearly far more practical than the long-winded, archaic names of Hevelius. In general, the placement and depiction of the larger topographic features of the lunar surface were in reasonable agreement in the two maps, but for many smaller or less prominent formations, there was less correlation. As the Riccioli map asserts that it is based on maps of Hevelius and others, with both corrections and augmentations, could not one assume that it was more reliable than its predecessors? How, in those days, could one judge which really was the more accurate without observing for one's self with a state-of-the-art telescope, a very rare and expensive item?

For a period of more than 150 years, the problem of choosing between these two major cartographic efforts for inclusion in textbooks and other related publications was usually circumvented by including copies of both the Riccioli and the 'Q' map, or sometimes the 'R' map with names added, of Hevelius. Because of the large number of names on each original map, plus the large dimensions of those maps, lists of names were often considerably shortened and map sizes cut down to fit into smaller text-book pages. To the extent possible, tables of comparative nomenclature were usually given even if only one map was included. However, this led to problems, since named formations on one map sometimes had no exact or even close counterparts on the other. Even worse situations arose when authors attempted to put Riccioli's names on copies of Hevelius's maps; in the last section of this chapter we will illustrate one of these incompatible marriages, and also note the name changes, movings, additions etc. that occurred prior to 1791, the date that ushered in a new era in selenography.

LUNAR OBSERVATIONS AND MAPPING, 1652–1790

Despite the differences between the Hevelius and Riccioli maps, their perceived excellence and scope, together with their quite wide-spread availability, inhibited any attempts at making a complete revision and update of them and their nomenclatures until the late 1780s, almost 140 years later, by which time their status as ultimate authorities was being questioned in earnest. However, this period did produce a mixed bag of interesting contributions to selenography, the most notable of which were: two important, pioneering Moon-mapping projects; one basic, simplified, annotated full-Moon image that, in the mid-1700s, rapidly superseded the Hevelius and Riccioli maps for inclusion in smaller-format books or where the Moon was a secondary subject; three other original Moon images or maps; one Moon globe; three cases of attention to the finer details of the lunar surface; and one case of blatant plagiarism! These are now discussed in a more or less chronological sequence, with a few textbook type maps included at the end for good measure.

1 Christopher Wren's globe

The famous architect of St Paul's Cathedral, the Octagon Room at Greenwich Observatory, the Royal Naval College at Greenwich, and many other fine buildings – Christopher Wren (1631–1723) – also deserves credit for being the first to construct a globe of the Moon. At the time (1661) he was a professor of astronomy at Oxford, and the globe had been requested by the recently formed Royal Society. Van Langren and Hevelius had both mentioned the advantages of a lunar globe, also one Matthias Hirzgarter (1574–1653), the frontispiece of whose 1643 book has a copy of Fontana's June 1630 image both upside down and reversed!

Wren's globe was based on a blank sphere 'fixed on a pedestal of lignum vitae' made by Joseph Moxon, a leading globe maker at the time, and it represented not only the spots and various degrees of whiteness upon the surface of the Moon, '. . . but the hills, eminences and cavities of it moulded in solid work, which if turned to the light shewed all the phases of the Moon, with the several appearances that arise from the shadows of the hills and vales.' Wren also apparently 'measured the relative positions of the different formations' himself. The globe was subsequently placed among the curiosities of the King's Cabinet. Wren also worked on the problem of the Moon's librations, apparently making some sort of model to demonstrate them. The globe was sold to a Mr Ary in 1749 and is now lost.

2 Montanari's quaint, unique map

Just over a decade after Riccioli's up-to-the-moment map became available, the Italian astronomer Geminiano Montanari (1633–87), better known for his discovery of the variability of the star Algol, composed a lunar map from a series of drawings of the Moon, from crescent to almost full, made on twelve consecutive clear nights in October 1662 (fig. 42). The portrayal of the shapes of the maria is quite crude, and their interiors are almost devoid of detail other than the major formations. The 'terrae', on the other hand, are crammed with detail, again rather crudely depicted, but comparison with modern maps or photos shows that most of the major features, as well as many of the apparently randomly placed squiggles, represent actual landforms. The engraving has a diameter of 38 cm. Being published in a rather obscure ephemeris and having no accompanying explanation or nomenclature, the map had no influence whatever on subsequent lunar cartography.

3 Hooke and Huygens scrutinize the details

The very few who were fortunate enough to have access to a large (e.g. 20 ft) telescope and observe the Moon through it cannot have failed to notice that although the Hevelius and Riccioli maps acted as quite good guides to the Moon's surface features, the depictions were highly stylized, and a large amount of interesting finer detail was totally missing. These shortcomings were first commented upon as early as 1664 by Robert Hooke (1635–1703), pioneer in several fields of physics and mechanics, who invited readers (of his *Micrographia*) to compare his drawing of the lunar formation Hipparchus (Mons Olympus in Hevelius's map) with the simple

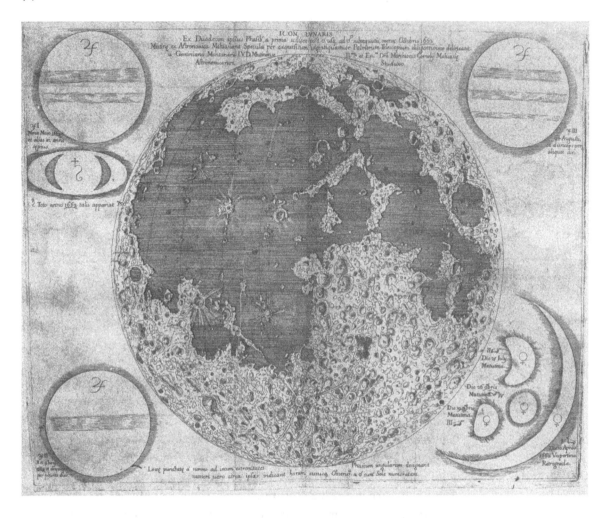

depictions of Hevelius and Riccioli (fig. 43). He also included quite graphic descriptions of this and other observed formations, with speculations on the nature of the surface and the origin of those formations, backed up by his own experiments with bullets being dropped into a slurry of pipeclay (impact theory), and bubbles of water vapour bursting from heated alabaster powder (volcanic theory).

Christiaan Huygens (1629–95), the famous Dutch scientist and pioneer also in many fields of physics and mechanics, who discovered Titan, Saturn's largest moon, and recognized that the mysterious appendages attached to Saturn were actually a flat ring, apparently observed the Moon on occasion. In 1658 he remarked on, but did not draw, '. . . quinque valles

Fig. 42. Geminiano Montanari's rather grotesque map from 1662. It was prepared after twelve consecutive clear nights of observation in that year.

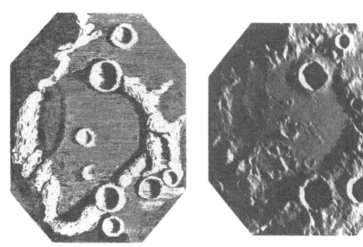

Fig. 43. Robert Hooke's 1664 drawing of the crater Hipparchus, with a modern photograph at similar phase for comparison. This is the first detailed drawing of a specific lunar feature.

exiguas rotundas ...', i.e. five bright, round vales, that he saw in the Mare Caspium of Hevelius. In 1685 he made a rough sketch (fig. 44a) of a feature that he described as a ditch (fossa) that proceeded from a brownish spot (macula) A, through a bright hollow (cavum) B, continung to C. The following year he drew a very rough sketch (fig. 44b) of another feature that he described as '. . . une fosse courbée irregulièrement aupres d'un rond enfoncé qi est fort clair.' – i.e. an irregularly curving ditch near a very bright, depressed circle. During the next lunation he observed and sketched, on two consecutive nights, an appearance that he had not seen before, namely, a long, very straight shadow resembling a sword (épée) (fig. 44c). He noted that the shadow was narrower on the second night, as should be expected.

With these last three observations he had discovered what are today named Rima Hyginus (Hyginus Rille), Rima Schröteri (Schröter's Valley), and Rupes Recta (The Straight Wall) respectively. He obviously gave these features special attention as being something different from the more common types of topography, and thus worthy of note. It is interesting to note that they were not rediscovered until over a century later, by Schröter. Since Huygens's observations were just notebook sketches and notes, they remained virtually unknown until their publication over two centuries later.

4 Chérubin and his cherubs

The French Capuchin friar Chérubin d'Orleans (1613–97, real name Michel Lasséré) published a large volume in 1671 on optics, in which, among other subjects, he describes his invention of a rhombic pantograph

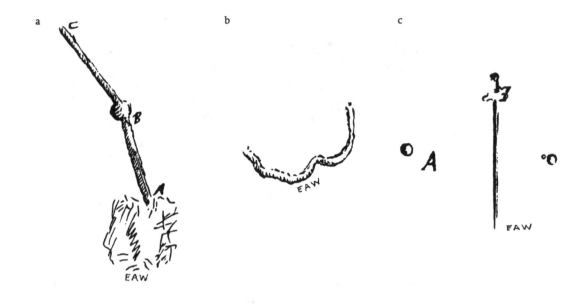

apparatus attached to a telescope and drawing board, by which accurate drawings of distant objects could be made. The point of the inner rhombus was located in the common focal plane of the objective lens and the eyepiece, and a pencil was placed in the joint of the outer rhombus. Thus with a stationary scene and a rigidly mounted telescope plus its accessories, it would be feasible to make enlarged drawings of distant scenes etc. For a moving object such as the Moon the apparatus would clearly be useless unless rigidly mounted to a sturdy equatorial mounting, something that was still to be invented!

It is thus a little surprising to find both a full-Moon image and a map with topography included in the book (figs. 45 and 46), with the following note above the former engraving:

> Observation of the lunar disk in its opposition to the Sun. Made by Father Chérubin d'Orleans Capuchin. By means of the instrument that he has recently invented, for proportionately drawing with exquisite exactitude, all types of objects, whether in the Sky or on the Earth, as they are viewed by means of the Dioptric Ocular.

The cherubs in the corner cartouches are depicted as using these instruments.

If the image and map look familiar, you are right. Compare them with Hevelius's 'P' and 'R' maps, and the similarities are plain to see. Thus Chérubin's claims (a) that the maps are from his own observations, and (b)

Fig. 44. Sketches of unusual features by Christiaan Huygens; (a) the crater Hyginus and rille; (b) the rille now known as Vallis Schröteri (Schröter's Valley); (c) the remarkable fault in Mare Nubium, now 'Rupes Recta' (The Straight Wall).

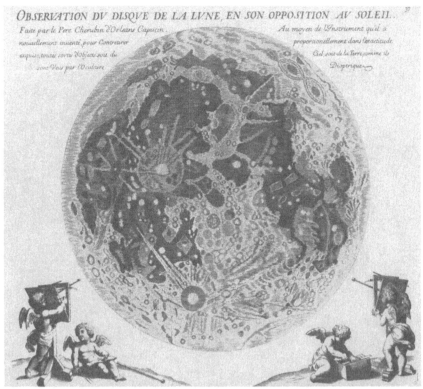

Fig. 45. Full-Moon image by Chérubin d'Orleans, supposedly made by using one of his inventions shown in the lower corners, but clearly copied directly from Hevelius's image.

Fig. 46. Another plagiarized map by Chérubin. The cartouche in the lower right corner shows a binocular telescope of his invention.

a

that he used his apparatus to produce them, are both quite false. Hevelius
rightly accused him of plagiarism.

5 Two coups for Cassini

The first of the major projects mentioned earlier was the production of a
detailed map of about 54 cm diameter, made at the Paris Observatory in

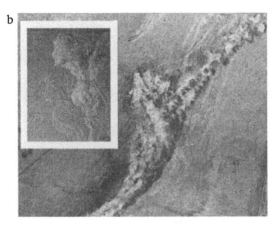

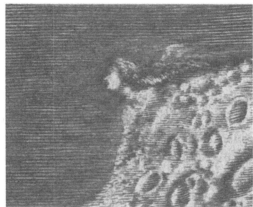

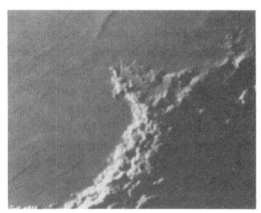

Fig. 47. (opposite and this page) The highly detailed and beautifully engraved, but very rare, map made under Cassini's direction and participation in 1679 (a). Much of the smaller detail is illusory. The lady's head at Prom. Heraclides is an interesting, whimsical feature of this map (b, top right); it was drawn this way (top left) in two of the original sketches. At the bottom is a modern photograph of this feature for comparison.

1679 by J.D. Cassini (1625–1712) and two able assistants, Leclerc and Patigny (fig. 47). A large number of quite detailed drawings of areas of the Moon observed at different phases were made during the 1670s, some 57 of which are still preserved at the Observatory. The final copperplate image has south up, but with the lunar axis rotated about 30° clockwise.

The large increase in amount of topographic detail portrayed is largely due to the increased size and quality of the telescopes used, but some of this detail is fictitious, and the positioning of the features and the outlines of the maria is generally inferior to that in the Riccioli map. However, the three-dimensional appearance imparted to the topographical features by the engraver (Patigny) is quite remarkable, and remained unsurpassed until much later. Very few prints were made from the copperplate, and no accompanying nomenclature or description was produced. Thus despite its large size, complexity and eye-catching appeal, it never became a serious challenger to the Hevelius/Riccioli monopoly.

If Cassini's large map was doomed to obscurity because of its rarity, then a much smaller image of the full Moon drawn up under his direction and published in 1692 in preparation for a total eclipse of the Moon predicted for July 28 of that year was destined for unprecedented prominence. The hope was that observers would view the eclipse and make accurate timings of the covering of a selection of lunar spots by the Earth's shadow, which could lead to improved longitude values for the observing sites. The chosen 40 spots were numbered in the approximate order in which they would be eclipsed on that particular occasion.

This image and its names table began to be copied, with ever decreasing accuracy and aesthetic appeal, from the 1750s to the early 1800s in encyclopedic dictionaries, French astronomy books, and especially in the annual astronomical almanac, *Connaissance des Temps*. Figure 48 shows the original copperplate image, but with the nomenclature added (*c.* 1733) in the corners and at the bottom. Some small nomenclatural changes that appeared in the original and were propagated through its numerous progeny are discussed a little later. A contemporary, related but esthetically more pleasing map was published in 1702 by P. de la Hire (1640–1718), being a reduction of his 13-ft-diameter image made in 1686 (fig. 49).

6 A second quaint seventeenth century image

An image of the full Moon, made in March 1694 by Georg Eimmart (1638–1705), a German engraver and amateur astronomer, is reproduced in Figure 50. Once again the boundaries of the maria have succumbed to artistic license, or perhaps just bad draughtsmanship. The whole image is completed with a distinctive but quite fanciful background 'filler'. As with the Montanari map, this image had no effect at all on subsequent lunar mapping. The diameter of the engraving is 28 cm.

7 Another closer look

In 1727, Francesco Bianchini (1662–1729) observed a limited area of the Moon with huge telescopes made by J. Campani. These instruments had lengths of about 30 and 50 ft respectively, and one can only wonder how anyone could manage to draw anything using such recalcitrant monsters suspended from masts and jiggling in the breeze. The image is of what Hevelius named the Alpes, plus Lacus Niger Major, Lacus Niger Minor, Mons Serrorum, and other formations with similar long-winded names. It

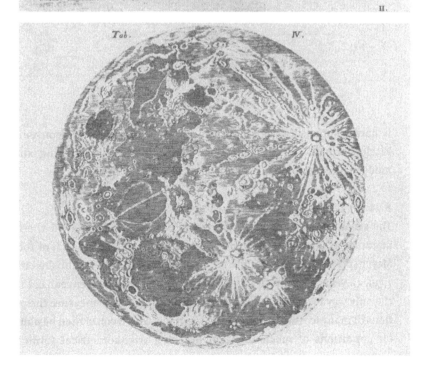

Fig. 48. Cassini's 1692 full-Moon image, prepared for an upcoming lunar eclipse. This formed the basis of numerous maps of ever decreasing accuracy and aesthetic appeal for over a century.

Fig. 49. Philippe de La Hire's full-Moon image from about 1702, showing an obvious kinship with that of Cassini, but with more detail.

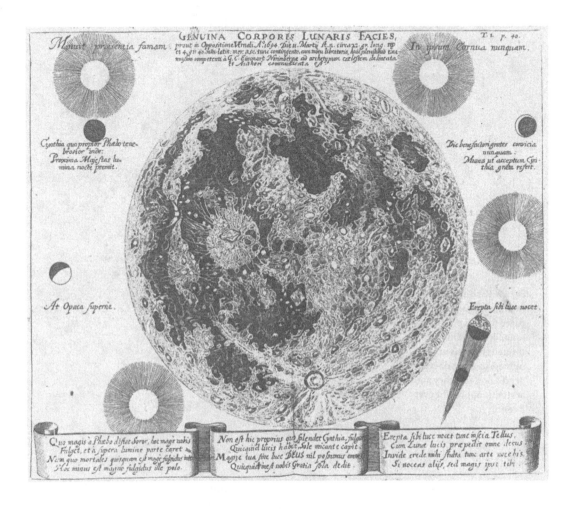

is hardly surprising that Bianchini chose to use the more convenient Riccioli names. An engraving made from Bianchini's drawing and a modern photograph for comparison are illustrated in fig. 51.

Fig. 50. A map by Georg Eimmart (1694) with notably contorted boundaries for the maria.

8 Mayer, pioneer of scientific selenography

The second important map referred to earlier, which marked the first major milestone in the accurate mapping of the Moon, was that of Tobias Mayer (1723–62). It was completed in about 1750 but not published until 1775, 13 years after his death, by Georg Lichtenberg. Mayer realized that the only way to construct an accurate lunar map was to measure the positions of many features on the Moon's surface, which could then be plotted for conditions of mean libration by using trigonometrical formulae.

a

b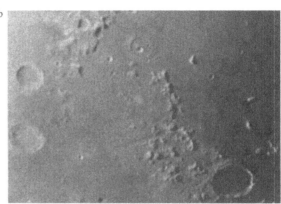

Fig. 51. 'Lunar spots with rectilinear shapes' as observed and drawn by Francesco Bianchini in 1727 (a). He shows the famous Alpine Valley (Vallis Alpes) for the first time, and the whole area can be seen to be quite well executed by comparing it with the modern photograph (b).

Filling in the intervening features from telescopic observations would then be far more accurate than the overall 'eyeballing' method used up to that time. Mayer measured the absolute positions of 23 well-marked lunar features, and made 40 drawings of various areas of the Moon as presented at different phases. From this material he drew up two manuscript maps, virtually identical except that one (at 40 cm) was twice the scale of the other. The 1775 engraved copy (20 cm diameter) of the smaller map is reproduced in fig. 52. Mayer intended to use the larger map as a basis to produce 12 gores for planned lunar globes. Six such gores were drawn, but the project was never completed.

Lichtenberg added the latitude–longitude grid to the smaller map and appended a list of 66 formations whose coordinates he had determined from their mapped positions, plus the 23 measured by Mayer at the telescope. He supplied both the Riccioli and Hevelius names for these features, but did not place their serial numbers on the map for fear of obscuring important details.

Comparing the map with a modern one or with lunar photographs shows that it is indeed more accurate than its predecessors, not only as regards topographic and shading detail, but also in the placement of that detail. The formations are drawn as observed under morning illumination.

9 Lambert's parallel effort

While Mayer's maps were lying aside unpublished and unknown, Johann H. Lambert (1728–77), more famous for his work on the scattering and absorption of light, mathematics etc., also conceived the idea of measur-

Fig. 52. The engraved version of Tobias Mayer's smaller map, which was drawn in 1749 and published in this form in 1775. It is the first map to be based on measured posi- tions of surface features, and is thus more accurate than its predecessors. The lines of latitude and longi- tude were added after Mayer's death by Georg Lichtenberg.

ing the absolute positions of a number of lunar features as an aid to drawing a map. He measured the positions of 66 formations and features, each to the nearest degree in latitude and longitude (compare with Mayer's measures, which are given to the nearest minute in each coordinate), tabulating them in the *Berliner Astronomisches Jahrbuch* for 1776. They are accompanied by a small image which has almost no scientific or artistic merit; a slightly later edition includes the names of 76 features, using Riccioli's nomenclature. (fig. 53) A yet later edition (fig. 54) uses a map projection that approaches Lambert's own azimuthal equal-area projection, but the 1775 publication of Mayer's map ensured the rapid demise of these images.

10 Copied maps

The period of 140 years covered in this chapter saw the publication of many astronomical textbooks, encyclopedias, dictionaries, ephemerides, atlases etc. that contained Moon maps. Some of these maps are aesthetically pleasing and artistic, while others are total abominations – for instance, compare the grossly degraded image in fig. 55 with its elegant progenitor, the 1692 Cassini image. The editor of the encyclopedia in which this appeared did not even notice that the image is reversed left-to-right! Two typical maps from this period that were not noted earlier in this chapter are presented in fig. 56, with relevant details given in the captions.

NOMENCLATURE CHANGES, 1652–1790

The numerous publications of this period that included all or selections of the two nomenclature schemes ostensibly followed the originals with no changes or additions. But mistakes always occur, and it is interesting to track these down and see whether or not they have been perpetuated to the present day. A few maps show new names not found in either Hevelius or Riccioli, while at least one has moved names to other formations – actions that are hardly conducive to a stable nomenclature! The Latin forms of the names used by both authors were retained almost exclusively throughout this period, but the list accompanying one edition of a French map (by La Hire, 1702; see fig. 49) containing 49 names from Riccioli, has them all in their French form.

1 Cassini makes three changes

The small annotated lunar image drawn up and published under J.D. Cassini's direction in 1692 (fig. 48) used Riccioli's names exclusively, a

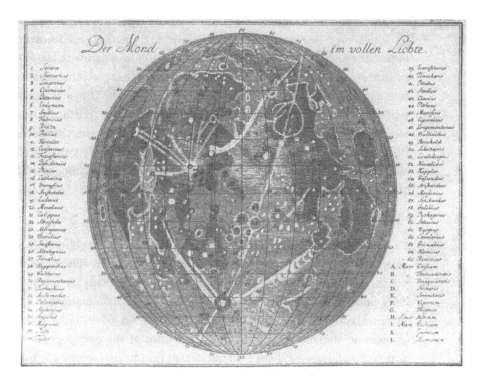

Fig. 53. A rather crude
map, presumably by
Johann Lambert, embrac-
ing idiosyncrasies of
Hevelius's and Cassini's
full-Moon images.

Fig. 54. Lambert's map
drawn on his equal-area
projection.

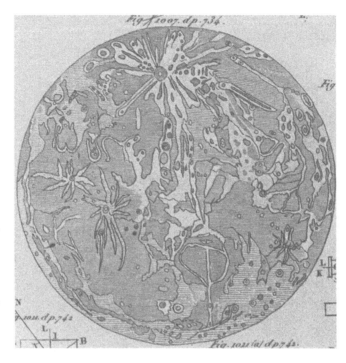

Fig. 55. Typical grossly degraded Moon map as found in encyclopedias of the late 1700s. The 'Greek phi' shape in Mare Serenitatis shows that it was based very loosely on the Cassini maps, but here we have a mirror image, exposing the carelessness of the engraver and inattention of the editor.

portent of the way things were destined to go 100 years ahead. Apart from a few small spelling errors, the following three deviations from the original Riccioli names are noted.

Insula Sinus Medii (Island of the Middle Bay)
> This appears as **Sinus Aestuum** (Bay of Seething Heat) in Riccioli. Van Langren has it as **Sinus Medius** (Middle Bay), but it is not known whether Cassini ever saw a copy of that map. The modern form **Sinus Medii** (Bay of the Center) originated with J. Bode in about 1780. In Hevelius it is **Mare Adriaticum**. Cassini's use of the term 'insula' is difficult to understand.

Promontorium Acutum (Sharp Promontory)
> Named **Prom. Methonis** (i.e. of Meton) by Van Langren, and **Prom. Heracleum** by Hevelius. Not named by Riccioli and not used after the beginning of the nineteenth century.

Promontorium Somnii (Promontory of a Dream)
> An unfortunate error. Riccioli has **Palus Somni** – with one 'i' only – which means Marsh of Sleep, situated not far from his **Lacus Somniorum**, Lake of Dreams. The **Palus** form prevailed, but the erroneous '**Somnii**' spelling was perpetuated until 1960.

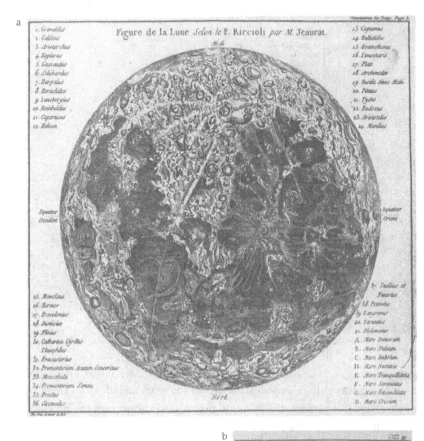

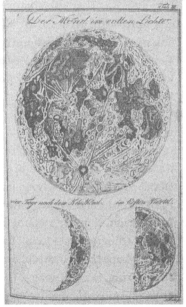

Fig. 56. Map from the 1770s, obviously based on the 1679 Cassini map (a), and another from the same era in a book by Johann Bode (b).

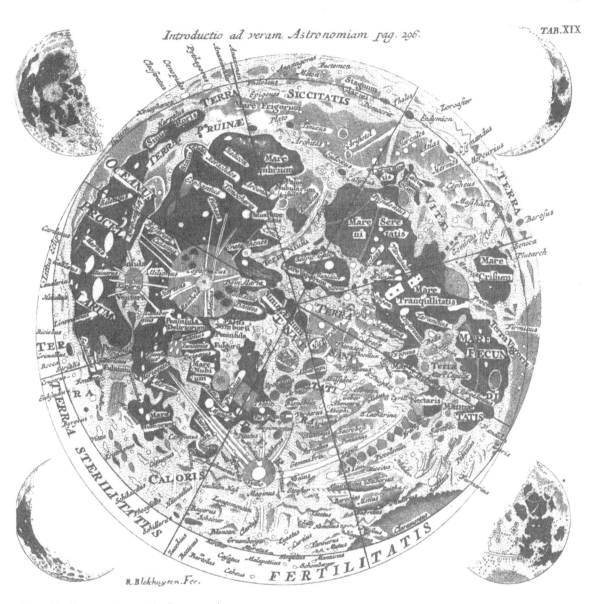

Fig. 57. Map from text-books by Keill in the early 1700s. This is clearly patterned after Hevelius's full-Moon image, but has Riccioli's nomenclature added – two incompatible sources. Try to find the origin of 'Goris' on the edge at the 4 o'clock position by comparing with fig. 39. Note also 'Flamstedius' sneaked in a short way above Mare Humorum, and several spelling errors (Mare Frigorum, Palus Nubularum, Sinus Aestium, Vitruinus, etc.).

Mac Platæ in Ayecluu Deliffine
ex Observ Illuftruu Blanchini

PLENILUNIUM NOVA FORMA ACCURATE EXPRESSUM CUM M
in suâ naturali apparenti magnitudine, situ, fig
collatis Tabulis Selenographicis Hevelii, Eustachij Divini, Gru
cum cælesti exemplari MELCHIOR A BRIGA

Mac Aristarchi
ex Eodem

1. Grimaldus   21. Tycho sup Longomont V
   sub Ricciolo   22. Eudoxus supra
2. Galilæus   Calippum
3. Aristarchus   23. Aristoteles
4. Keplerus   24. Manilius cum Sociis
5. Gassendus   25. Menelaus
6. Schikardus   26. Hermes
7. Harpalus ad   27. Possidonius
   sinum Roris   28. S. Dionysius Areop.
8. Heraclides ad   29. Plinius cu Soc
   sinum Iridum   30. S. Catharina
9. Lansbergius   31. Fracastorius
10. Reinholdus   32. Promontor.
11. Copernicus cuSoc: acutum
12. Helicon   33. Messala
13. Capuanus   34. Promontor.
    et Cicchus   somnii
14. Bullialdus   35. Proclus
15. Eratosthenes   36. Cleomedes
16. Timocharis   37. Snellius, et
17. Plato cum Sociis   Furnerius
18. Archimedes   38. Petavius
19. Insula sinûs   39. Langrenus
    medii   supra Wendelinum
20. Pitatus   40. Taruntius
☉ Meton. ♀ Prosatius ♆ Tres sBReges Magi
Regio Lunaris circa mac Platonis
ex Observat ejusdem Blanchini

Aristarct et Calippus   Riccioli
Blanchino nuncuminata

Scala partium æqualium, et Diametri

Ferdus Soarianus delin.   Variantur ex motu anomaliæ apparens magnitudo, ex oscillatio

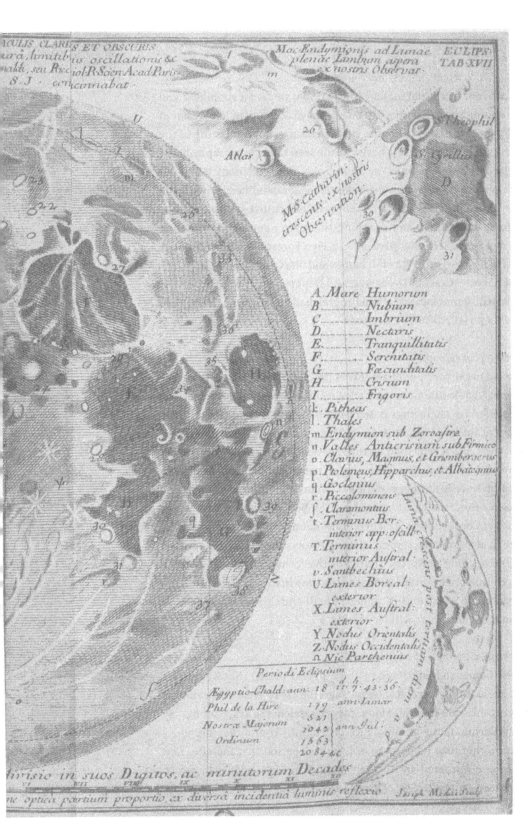

Fig. 58. Map by Melchior à Briga (1747). Note three starlike bright spots just below the centre, identified by 'psi', which are named 'Tres S.S.Reges Magi'; and 'Nic Parthenius', identified by a capital 'omega'. These names were never used subsequently.

2 *Allard adds, subtracts, moves, and changes names*

In one corner of a celestial map by Allard (1700) we find a simplified and much reduced copy of Divini's full-Moon image accompanied by what at first sight appears to be the Cassini nomenclature and numbering of the 40 craters plus eight maria. We have already seen that a topography-based nomenclature cannot be put onto an image of the full Moon; Allard not only attempts to do this, but also takes considerable liberties with the names by replacing twelve with different names culled at random from the original Riccioli list, and dropping the mare names and renaming them **Prima Sylva**, **Secunda Sylva** etc. (i.e. First Forest, Second Forest etc.). He also replaced two other Riccioli names and one Cassini name with new personal names – **Cassini**, **Descartes**, and **Neperus** all of which are used today but in different locations; Van Langren had already used the last two of these, but on still different locations.

3 *Keill sets a poor example*

I mentioned in the previous chapter that problems arose when authors of astronomical books attempted to apply Riccioli's names to the formations as represented on engravings based on Hevelius's 'R' map (fig. 34). The worst case of this type can be seen in fig. 57, which is taken from a very respected and much reprinted astronomy textbook by Keill (first half of the eighteenth century), where the Riccioli names are found on an engraving based on Hevelius's image of the full Moon ('P', fig. 33) – the very one that does *not* show the topographic features! Because of this, very few of the names can be seen to be attached to any definite features.

This map also contains several spelling errors, and I leave readers the exercise of tracking down the source of the formation 'Goris' at the 4 o'clock position on the limb. Halfway to the limb in the 8 o'clock direction we find the name '**Flamstedius**', added by Keill. Did this sneaky move have any permanent effect? We will see shortly.

4 *A few further attempts at additions*

Some new names that appeared in this 140-year period that were never adopted are as follows: **Tres S.S.Reges Magi** (the Three Wise Men), and **Nic Parthenius**, on a rare 1747 full-Moon image by Melchior à Briga (see fig. 58); and **Mons Carmaniae** on a copy of Riccioli's map by Le Monnier (fig. 59).

The only other map I am aware of from this period that includes new names is one authored by Father Maximilian Hell (1720–92), which appears in early copies (1760s) of his *Ephemerides*. This map (fig. 60) is a poor

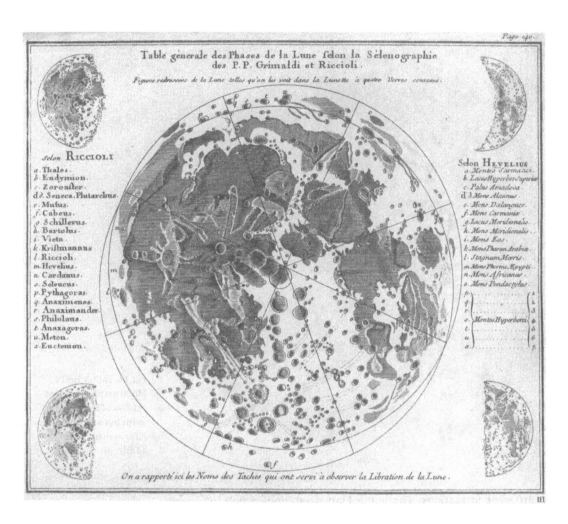

Page 140.

Table generale des Phases de la Lune selon la Sélénographie des P.P. Grimaldi et Riccioli.

Figures redressées de la Lune telles qu'on les voit dans la Lunette à quatre Verres convexes.

Selon RICCIOLI

a. Thales.
b. Endymion.
c. Zoroaster.
dd. Seneca, Plutarchus.
e. Mutus.
f. Cabeus.
g. Schillerus.
h. Bartolus.
i. Vieta.
k. Kristmannus.
l. Riccioli.
m. Hevelius.
n. Cardanus.
o. Seleucus.
p. Pythagoras.
q. Anaximenes.
r. Anaximander.
s. Philolaus.
t. Anaxagoras.
u. Meton.
x. Euctemon.

Selon HEVELIUS
a. Montes Sarmatici.
b. Lacus Hyperboreus superior.
c. Palus Amadoca.
d. Mons Alaunus.
e. Mons Dalangner.
f. Mons Carmaniæ.
g. Lacus Meridionalis.
h. Mons Meridionalis.
i. Mons Eos.
k. Mons Pharan Arabiæ.
l. Stagnum Morris.
m. Mons Thermæ Ægypti.
n. Mons Africanus.
o. Mons Pondactylus.

On a rapporté ici les Noms des Taches qui ont servi à observer la Libration de la Lune.

Fig. 59. An excellent reduced copy of the Grimaldi/Riccioli map from Le Monnier's *Institutions Astronomiques* (1746). 'Mons Carmaniae' (f), is supposedly from Hevelius, but he does not list the name anywhere.

copy of one that appeared a few years earlier in Homann's atlas under the authorship of Doppelmaier, which is itself a poorish copy of the Riccioli map. Here are the 11 names added by Hell.

Flamsteedius	**Halleyius**	**Malebranchius**	**Regnaultius** S.J.
Rostius	**Scharpius**	**Schmelzerus** S.J.	**Schottus** S.J.
Tacquettus S.J.	**Volsius** S.J.	**Wolffius Freiherr**	

Most of these apply to nondescript bright patches or indefinite features, and the identifications were never accepted except for **Scharpius**, a germanized latinization of the English name **Sharp**, who was Flamsteed's assistant at the Royal Observatory, Greenwich. The names **Halley**, **Rost**, **Tacquet**, and **Wolff**, no longer in their latinized forms, were later placed

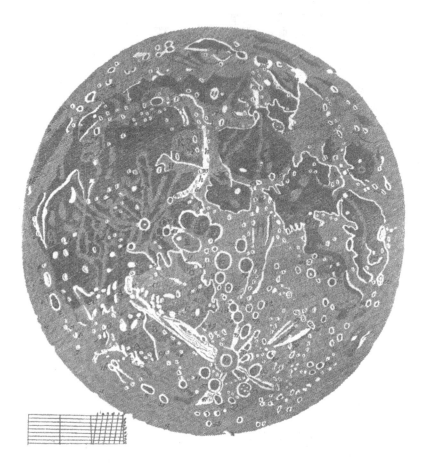

Fig. 60. Father Hell's
Moon map, a degraded
version of Hevelius's
maps but with many fan-
ciful and non-existent
additions.

on different formations. The name **Flamsteed** rings a bell, of course,
having been added by Keill some 60 years earlier. Hell's formation is near
Keill's, but is not the same, and neither of these is the present day
Flamsteed, which, however, lies in the same general area. As noted earlier,
J. Bode changed Cassini's '**Insula Sinus Medii**' to '**Sinus Medii**' in about
1780, an apparently logical improvement since this dark area is certainly
not an island in the lunar sense. Lichtenberg applied names to two features whose positions had been measured by Mayer, but which were
unnamed on the Riccioli map. These are the crater **Mercurius falsus** (nowadays 'Gauss'), located near Riccioli's 'Mercurius', and the promontory
Heraclides falsus (nowadays 'Prom. Laplace'), lying across Sinus Iridum
from Riccioli's 'Heraclides'.

I have purposely belaboured the point about spelling and identification
errors, and the adding or moving of names, because similar actions have
caused much confusion and unnecessary work in more recent times.

CHAPTER 6

....

A GLOBE, TREE RINGS, AND A CITY

TIME TO MOVE AHEAD

As already noted, right from the time of its publication up to the last decade of the eighteenth century – a period of 140 years – Hevelius's *Selenographia* remained the most comprehensive reference source available that dealt with the formations on the lunar surface. Riccioli's simple nomenclature scheme usually took precedence over that of Hevelius, while his map enjoyed about equal popularity, but his rather abbreviated descriptive text ensured an overall preference for Hevelius.

The extremely limited production of Cassini's large, detailed map of 1679, plus the fact that it had no nomenclature or descriptive material accompanying it, meant that this map was never in contention as a prime reference during this period. His 1692 full-Moon image, although much reproduced (and degraded) in the 1750–1800 era, was unsuitable as an illustration for any text that dealt in detail with the lunar surface topography. Similarly, the Tobias Mayer map, even though it ushered in the age of true lunar cartography and was more accurate than anything that had preceded it, was not accompanied by descriptions of the lunar surface features; it was still a relatively new production in the 1780s, and apparently not widely available. The end of *Selenographia*'s reign of pre-eminence was imminent, however, mainly because of some notable advances in optics.

IMPROVED TELESCOPES MAKE THEIR DEBUT

Towards the end of the eighteenth century, both reflecting and refracting telescopes were increasing in availability, quality, and manageability. Would-be Moon observers, armed with these much improved and more

Fig. 61. A typical textbook
Moon map (late 1700s),
the type that motivated
John Russell to prepare
his maps and globe.

convenient (i.e. 'user-friendly', to use the current parlance) instruments,
cannot have failed to notice that the commonly available maps, such as
the degraded copies of the 1692 Cassini or much-reduced Hevelius 'P'
images, acted as very poor guides to the topographical features. Even in
the original Hevelius or Riccioli maps, if one were fortunate enough to
have one available, the depictions of that topography – as already commented upon by Hooke – fell far short of reality, and a large amount of
interesting intervening detail was totally missing.

RUSSELL GOES IN ONE DIRECTION . . .

John Russell (1745–1806), an English artist and Royal Academician of some
note, was struck by the beauty and wealth of detail seen in a telescopic
view of the Moon. Thus in a letter dated 9 February 1789 to the Radcliffe
Observatory in Oxford he writes, '. . . how much struck a young Man conversant with Light, and Shade, must be with the Moon in this (near first
quarter) state; especially, as I was not taught to expect such clearness and
expression, as is to be found near and upon the indented Edge; . . .'. Being
conversant at that time with only the 'very inferior Prints to be met with
in common Dictionarys', (e.g. fig. 61), 'such unsatisfactory imitations,

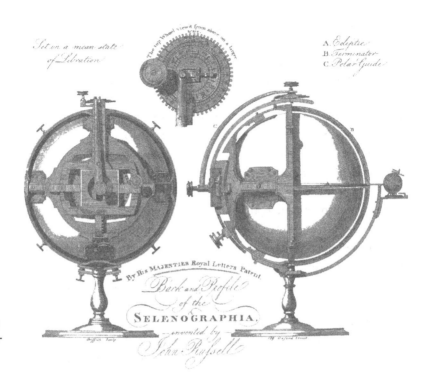

Fig. 62. Contemporary (1797) diagrams showing the mechanical construction of Russell's 'Selenographia'.

both as to incorrectness of Form and Effect led me to conclude I could produce a drawing in some measure corresponding to the feelings I had upon the first sight of the gibbous Moon through a telescope . . .', he was somewhat taken aback on first 'meeting with' the coloured copies, by Doppelmayer, of Hevelius's and Riccioli's maps in the Homann atlas. He says he 'was again at a stand' after obtaining a copy of Hevelius's *Selenographia*, but that he thought he could produce results that were more artistic.

A little later he was shown a copy of the large Cassini map, which again impressed him, but he was fully aware of its shortcomings and was determined to press ahead and produce his own map. While making pencil sketches of all visible areas of the Moon, a program that he commenced as early as 1764 and continued for 40 years, he also measured the relative positions of 34 prominent features. The first product of this effort was not a flat map (planisphere) but a series of gores, engraved by Russell himself, to be pasted onto globes, each 12 inches in diameter. Of the few (about seven) globes that were made, five or six were mounted in a complicated brass mechanism by which the lunar librations, tilt of the lunar axis, and the sunrise or sunset line could be demonstrated (fig. 62). The globes, named 'Selenographias' by Russell, are dated 1797. Nine years later, he published two lunar images, one of full Moon and the other with topographic details added (figs. 63 and 64), both about 36 cm in diameter.

The former is clearly far more detailed than any of its predecessors, and the very complex interplay of delicate shadings reveals the hand of a master artist. Indeed, the highly detailed nature and general accuracy of this image have never been surpassed. Of course, the advent of photography later in the nineteenth century soon discouraged any further attempts at such an exacting and laborious task. Some of the surface topography portrayed in the second image is perhaps a little less reliable both in its positioning and reality.

Russell never placed a coordinate grid of lunar latitudes and longitudes on his images, nor were they accompanied by any form of nomenclature. Add to these disadvantages the fact that very few copies of the planispheres were printed, together with the restricted production of the globes, and one can understand why Russell's work never had the impact in selenographical circles that its artistic and scientific contributions merited. Figure 65 illustrates the same lunar area as shown in fig. 41, but here depicted by (a) Cassini (1679), (b) Mayer (1749), and Russell (1805).

. . . AND SCHRÖTER GOES IN ANOTHER

In about 1787, well after Russell initiated his observational program, Johann Hieronymus Schröter (1745–1816), chief magistrate in Lilienthal near Breman, Germany, and keen amateur astronomer, started up his own ambitious program – to observe and delineate lunar features under all conditions of illumination; to measure the heights and depths of the more important elevations and craters respectively; to look for evidence of a lunar atmosphere or changes on the lunar surface; and finally to make a map 46.5 inches in diameter, based on Tobias Mayer's measures. He had observed the Moon on occasion over the previous three years, but mainly to note the appearance of the crater Aristarchus when not illuminated by the Sun, following William Herschel's announcement in 1783 that it was glowing like a volcano.

He obtained a 2.5 inch refractor by Dollond, and 4.75 and 6 inch reflectors by W. Herschel; later, he added a 9.5 inch reflector by Schrader, and finally constructed an 18.5 inch reflector, a large clumsy and not very efficient monster. To aid in the drawing, he invented his 'Projections-Maschine', a simple contraption of somewhat doubtful efficacy whereby he viewed the image of the Moon's surface through the eyepiece using one

Fig. 63. Russell's
engraved image of the
full Moon, 1805.

Fig. 64. Russell's map
with topography
included.

eye, making a drawing on a board fixed to the telescope while using the
other eye. A rotatable glass reticle with a grid of small squares was situated
in the image plane, and the drawing paper was placed on the board, also
rotatable. The paper was pre-marked with dots forming half-inch squares.
In use, the reticle was rotated until one set of lines paralleled the line
between the Moon's horns, and the board then adjusted in angle and dis-
tance until the two grids were congruent – obviously not a very precise
operation!

Schröter observed the Moon diligently, amassing numerous drawings of
various areas of the disc, measuring mountain heights and crater depths

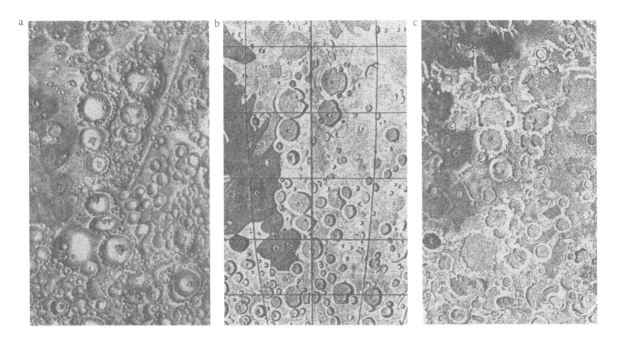

Fig. 65. A portion of the maps of (a) Cassini (1679), (b) Tobias Mayer (1749), and (c) Russell (1805). Compare these with the four maps shown in fig. 41, which are of the very same area.

Fig. 66. (overleaf) An engraving of Tobias Mayer's map, here with south up, as it appears in Schröter's first volume of observations (1791). The list includes names from both Hevelius and Riccioli.

from shadow lengths, searching for evidence of changes on the surface or the existence of an atmosphere, etc., so that by the end of 1790, he had enough material to publish in the form of a large volume. He gave this the title of *Selenographische Fragmente*, very apt in view of the fragmented nature of the subject matter and areas dealt with. This book was published in 1791, and further similar observational work appeared as volume 2 in 1802.

As a guide for readers, he included a new engraving of Mayer's map, but with numbers placed on the 89 features listed by Lichtenberg and letters on 18 of Riccioli's 21 'watery' designations. A list of these 107 identified features surrounds the map, with both Hevelius and Riccioli names given (fig. 66). The engraving was made by Georg Tischbein, a Bremen artist, who also engraved all the numerous illustrations for the two volumes. It has sometimes been stated that he was a mediocre artist, judging by the rather naïve appearance of the drawings (e.g. figs. 67 and 68a), but his rendering of the Mayer map (cf. fig. 52) shows that this is an unjust criticism. We conclude that Schröter's artistic talents were somewhat limited (fig. 68b is one of his best). Note that the map is oriented south up, as viewed in most astronomical telescopes, and thus follows the trend initiated by Fontana and carried on by Rheita, Cassini, La Hire and Bianchini.

Nomina macularum insigniorum
sec. Ricciolum,   *sec. Hevelium.*

TOB. MAYERI TABU

	sec. Ricciolum	sec. Hevelium
A,	Mare Crisium	Palus Maeotis
B,	M. Foecunditatis	Mare Caspium
C,	M. Nectaris	Sin. Athen. et Sin. extr Ponti
D,	M. Tranquillitatis }	Pontus Euxin.
E,	M. Serenitatis }	
F,	Lacus Somniorum	Sinus Cercinites
G,	Lac mortis	Montes Peuce
H,	Palus Somnii	Lac. Corocondametis
J,	Mare Frigoris	Mare Hyperboreum
K,	M. vaporum	Propontis
L,	Sin. aestuum	Mare Adriaticum
M,	Mare nubium	M. Pamphilium
N,	M. humorum	Sin. Sirbon. et M. Aegyptiac.
O,	Sinus epidemiarum	Insula Didymae
P,	Oceanus procellarum	Mare Coum et M. medit. pars
Q,	Mare imbrium	Mar. mediterr pars septent.
R,	Sinus iridum	Sinus Apollonis
S,	Sinus roris	Sinus Hyperboreus

1	Seneca	Mons Alaunus
2	Mercurius falfus	
3	Mercurius	Lacus hyperboreus inf.
4	Langrenus	Insula maior
5	Vendelinus	
6	Furnerius	Pars montis Paropamifi
7	Cleomedes	Pars mont. Riphaeor.
8	Petavius	Petra Sogdiana
9	Steinius	P. mont. Paropamifi
10	Endymion	Lac hyperber. Sup.
11	Snellius	M. Paropamifus
12	Taruntius	Sin. Phasianus
13	Atlas	Pars M. M. Macrocemn.
14	Proclus	M. Corax
15	Goclenius	M. Caucafus
16	Hercules	P. Mont. Macrocemn.
17	Cerforinus	P. Mont. Herculis
18	Eracaftorius	Lac. Thospitis
19	Piccolominius	Pars M. M. Sogdian
20	Poffidonius	Insula Macra
21	Vitruvius	Apollonia maior
22	Theophilus	Pars M. Moschi
23	Cyrillus	Pars M. Moschi
24	Plinius	Promont. Archerufia
25	Catharina	Pars M. Moschi
26	Dionyfius	Pars M. Herminii
27	Ariftoteles	M. Serrorum
28	Eudoxus	M. Carpathes
29	Menelaus	Byzantium
30	Calippus	M. Aemus
31	Maurolycus	
32	Abilfeda	Pars M. Antitauri
33	Manilius	Insula Besbicus
34	Apianus	Pars Anti–Libani
35	Stoeff erus	M. Calchaftan

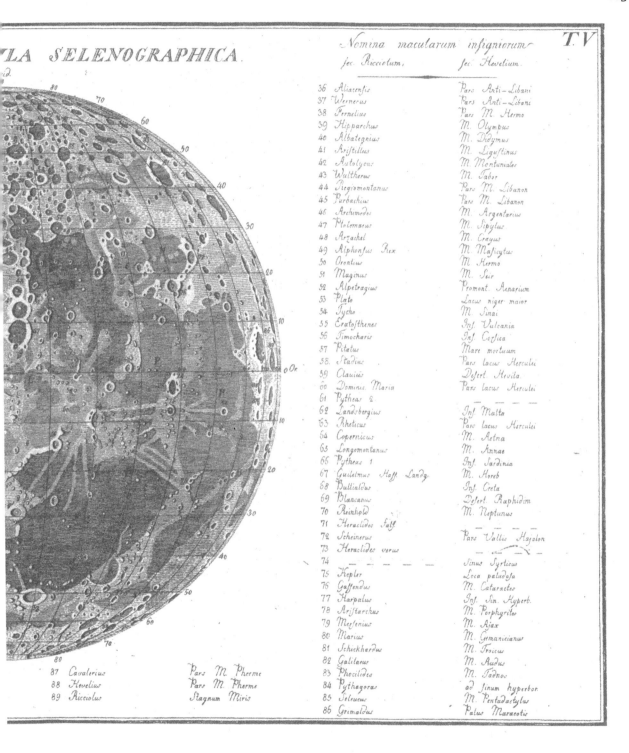

LA SELENOGRAPHICA.

T.V

Nomina macularum insigniorum
sec. Ricciolum.   sec. Hevelium.

36	Aliacensis	Pars Anti–Libani
37	Wernerus	Pars Anti–Libani
38	Fernelius	Pars M. Hermo
39	Hipparchus	M. Olympus
40	Albategnius	M. Didymus
41	Aristillus	M. Ligustinus
42	Autolycus	M. Montaniales
43	Waltherus	M. Tabor
44	Regiomontanus	Pars M. Libanon
45	Purbachius	Pars M. Libanon
46	Archimedes	M. Argentarius
47	Ptolemaeus	M. Sipylus
48	Arzachel	M. Cragus
49	Alphonsus Rex	M. Masicytus
50	Orontius	M. Hermo
51	Maginus	M. Seir
52	Alpetragius	Promont. Aenarium
53	Plato	Lacus niger maior
54	Tycho	M. Sinai
55	Eratosthenes	Inf. Vulcania
56	Timocharis	Inf. Corsica
57	Pitatus	Mare mortuum
58	Stadius	Pars lacus Herculei
59	Clauius	Desert. Hevila
60	Dominus Maria	Pars lacus Herculei
61	Pytheas 2	
62	Landsbergius	Inf. Malta
63	Rheticus	Pars lacus Herculei
64	Copernicus	M. Aetna
65	Longomontanus	M. Annae
66	Pytheas 1	Inf. Sardinia
67	Guilelmus Haff. Landg.	M. Horeb
68	Bullialdus	Inf. Creta
69	Blancanus	Desert. Raphidim
70	Reinhold	M. Neptunus
71	Heraclides falf.	
72	Scheinerus	Pars Vallis Hajalon
73	Heraclides verus	
74	— — — —	Sinus Syrticus
75	Kepler	Loca paludosa
76	Gassendus	M. Cataractes
77	Harpalus	Inf. Sin. Hyperb.
78	Aristarchus	M. Porphyrites
79	Messenius	M. Ajax
80	Marius	M. Germanicianus
81	Schickhardus	M. Troicus
82	Galilaeus	M. Audus
83	Phocilides	M. Taenos
84	Pythagoras	ad Sinum hyperbor.
85	Seleucus	M. Pentadactylus
86	Grimaldus	Palus Maraeotis

87	Cavalerius	Pars M. Pherme
88	Hevelius	Pars M. Pherme
89	Ricciolus	Stagnum Miris

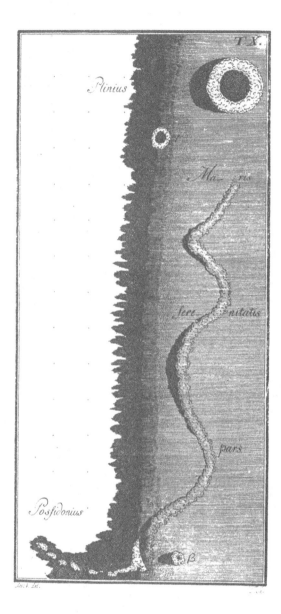

Fig. 67. One of Schröter's earliest sketches, showing the serpentine ridge in Mare Serenitatis in rather naive fashion.

SCHRÖTER'S NOMENCLATURE

The average scale of the majority of Schröter's drawings as published is about 1:3 million, which corresponds to a lunar diameter of around 4 ft – that of his proposed complete lunar map. At this scale, only one or two named formations appear on any given drawing, quite insufficient when describing the many new features included. Thus Schröter had to make a decision on how to name or otherwise tag these newly recorded features so

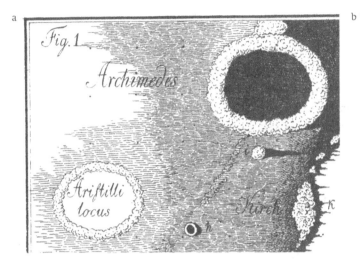

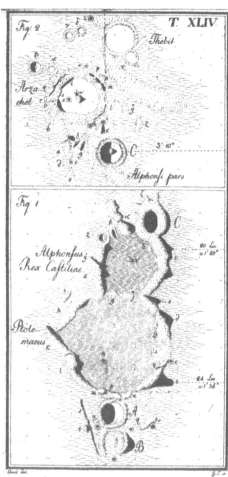

Fig. 68. A portion of a later drawing by Schröter showing his persistent depiction of sharp crater rims as a ring of trees (a), and (b) two of his later drawings that are esthetically quite pleasing, reasonably accurate, and notably more detailed than anything produced by earlier observers.

that they could readily be identified from his written descriptions. But this was far from being the only problem; the Moon was now known to have no water on its surface, yet no less than 13 of Hevelius's general categories have watery connections such as Chersonnesus (peninsula), Fluvius (river), Stagnum (swamp) etc., which are thus quite inappropriate. Riccioli has six such categories – more manageable but still not appropriate. And yet the most prominent topographic features on the Moon's surface, the circular depressions and mountain rings, had received no general name whatever!

Add to these problems the fact that Latin had been almost entirely phased out as the international language towards the end of the eighteenth century, books then being written in the language of their country of origin, we can appreciate that Schröter was faced with somewhat of a dilemma. The Latin names should be preserved for continuity and because of their long usage, yet he wished to write the text for his book in his native German, with German names for the general landforms seen on the Moon. Thus he gives such descriptive words as **Bergketten** (mountain chains), **Bergadern** (ridges), **Thalern** (valleys), **Flachen** (plains), **Ringebene** (ring plains), **Wallebene** (walled plains), **Ringgebirge** (ring montains), **Einsenkungen** (depressions), and several others of a similar sort.

We also find for the first time, to the best of my knowledge, repeated use of the words **Rille** and **Crater**. The former is the German word for a groove, a perfect description for these features, with no implication regarding their mode of formation. We will see later how incorrect translations into English led to quite erroneous conceptions about their nature. The latter word is of course not German in origin, but Latin, and generally signifies a large shallow bowl. Actually, the word was used by Allard in 1700 in a short note (in Latin) under his small map, in which he remarks that the pits (putei) that cover the greater part of the Moon are comparatively deeper than water bowls (Crateres aquatici), and many are considered to have central, rounded hills (collis rotundi). Since Schröter does not mention Allard at all, it is probably fair to say that Schröter was responsible for introducing this now universally used word into selenography in particular and planetary studies in general.

RICCIOLI GETS THE NOD

Although Hevelius's cumbersome names had led to their almost total elimination from such publications as encyclopedias and dictionaries, Schröter correctly decided to include them in his far more comprehensive work to act as a link to Hevelius's pioneering *Selenographia*. Thus for all those formations that are included in his drawings, he gives both Riccioli and Hevelius names in the text and index. However, except for the single case of Riccioli's '**Terra Nivium**' (Land of Snows) where Schröter chose Hevelius's more appropriate '**Montes Apenninus**', only the Riccioli names are given on the actual drawings. This was a wise move, not only because of their shortness but also because he wished to add many new names of his own choice. Trying to find more names from ancient terrestrial geography and maps would have been a daunting task, but there was no trouble in finding names of deserving scientists, mathematicians etc. Yet another consideration was the fact, noted earlier, that Hevelius sometimes gave a single name to a whole group of craters, a definite disadvantage when trying to describe the separate components of the group.

SEVENTYSIX NEW NAMES, AND LETTERS GALORE

Referring to fig. 69, which is typical of Schröter's full-page drawings, we find four of Riccioli's regional names (**Mare Imbrium, Sinus Iridum, Sinus Roris**, and **Terra Pruinae**), but only three names for topographical features (**Harpalus, Heraclides**, and **Helicon**). Schröter has added numerous details not shown in Riccioli's map, and has supplied new names for the craters **de la Condamine, Maupertuis**, and **Bianchini**. Also added is **Scharpius**, obviously taken from Hell's map, and **Heraclides falsus**, from Lichtenberg's nomenclature table for his publication of the Tobias Mayer map.

Looking now at the letters, there are at least 60 of these designating an assortment of topographic features. You will find upper and lower case script letters, upper case Roman letters, and lower case Greek letters – all used willy-nilly with no apparent rules for either their choice or distribution. Elsewhere, the same formation appearing on different drawings may have different letters.

Another problem that confronted Schröter was that of correlating some of the named formations in the Riccioli map with those in his drawings. Even nowadays, with the availability of dozens of whole-disc photos of the Moon, it is quite an exercise to identify some of the Riccioli formations; it must have been far more difficult for Schröter, and it is hardly surprising that he misidentified a few. A detailed review of Schröter's nomenclature is given in Appendix H.

REVIEW OF SCHRÖTER'S DRAWINGS

The 75 engraved plates published in the two volumes include anything from whole-page drawings of larger areas to groups of twelve or more sketches of specific small details. Examination shows that while a few drawings appear quite amateurish (fig. 66), others are reasonably accurate in their portrayal. Schröter consistently gives the rims of craters the appearance of an overhead view of a ring of closely spaced trees (fig. 68a), even though many of those craters display sharp rims as viewed in the telescope. Nevertheless, comparing the many drawings with modern photos shows that they include virtually all of the more important details of each region except in only one or two rare cases where he apparently became confused by what he observed.

Whatever criticisms may be leveled against Schröter's work, it can fairly be said that he pioneered the science of detailed and comprehensive selenography which, with Mayer's pioneering attention to positional accu-

T. XXIV

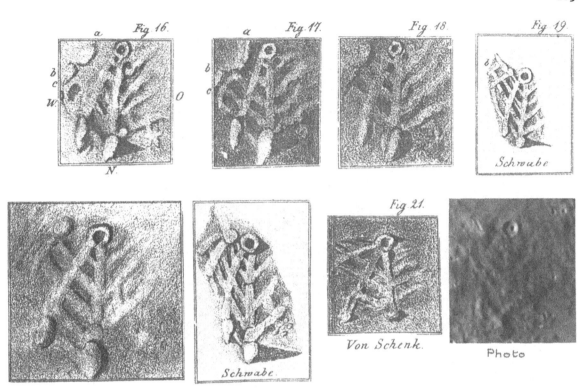

Fig. 69. (opposite) Another typical 'fragment' by Schröter showing some of his newly added names (Maupertuis, de la Condamine, and Scharpius – the latter being taken from the earlier map of Maximilian Hell. Note the numerous Roman and Greek letters identifying various features of interest.

Fig. 70. (above) Gruithuisen's famous – or infamous – lunar city. Four sketches are his, plus two by Schwabe and one by Von Schenk. The photo at bottom right shows the actual appearance.

racy, laid the ground for an unprecedented burst of lunar observation and cartography in Germany.

GRUITHUISEN INVITES SOME RIDICULE

The painstaking but incomplete selenographic work of Schröter undoubtedly acted as an incentive for others to pick up where he had left off, or at least to take an interest in lunar observation. One such person was Franz von Paula Gruithuisen (1774–1852), a native of Bavaria who began his professional career in the field of biology and medicine, and invented a surgical device for crushing bladder stones. He was later appointed professor of astronomy in Munich. His publications on lunar and related matters began in 1821, and appeared in a variety of rather obscure journals and handbooks for some 28 years. He concentrated on observing and drawing quite limited areas of the lunar surface, and it was here that his imagination ran wild. He thought that he could detect evidence of lunar inhabitants, such as huge buildings, waterways etc. The one item that drew the most attention, however, was his supposed lunar city. It also brought ridicule from the astronomical community, who pooh-poohed such fantastic notions.

Fig. 71. For readers who have telescopes, a photo outlining the 'city', which lies between the craters Mösting (top centre) and Eratosthenes (bottom right).

However, it did incite other observers to examine the supposed city for themselves, and show that it was a natural formation, albeit apparently unique on the Moon's visible surface. Figure 70 shows a group of sketches of the formation by Gruithuisen and two other observers, together with a photo for comparison. It should perhaps be noted that he was using a small refractor telescope, 2 inches in diameter actually, which is scarcely enough to resolve the two sets of sub-linear features of the 'city' into their separate, knobbly elements. A similar situation arose later in the nineteenth century, when what appeared to be narrow, dusky linear features were first seen on the planet Mars. Subsequent telescopic observations, and later on, photos from various spacecraft, showed that these were what had been suspected for many years – sub-linear arrays of darker surface markings. For readers who might like to observe this feature for themselves, it is situated in the dark area about halfway between the craters Mösting and Eratosthenes (fig. 71).

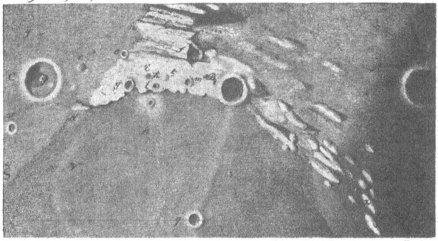

Gebirge zwischen Plinius und den Apenninen.

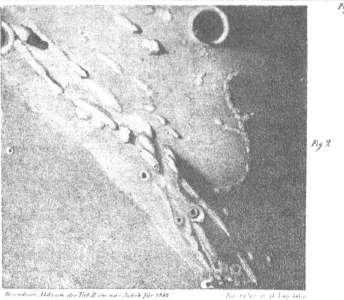

Fig. 72. Two of Gruithuisen's lithographs of the SW area of M. Serenitatis.

DRAWINGS AND A MAP

Gruithuisen published, besides the mostly small sketches of lunar areas and details (fig. 72), for many of which he produced his own lithographs, an aesthetically pleasing lithograph map of the whole Moon (1825, fig. 73), which is essentially Mayer's map but with a few minor modifications. This version had no nomenclature appended, but another apparently very rare

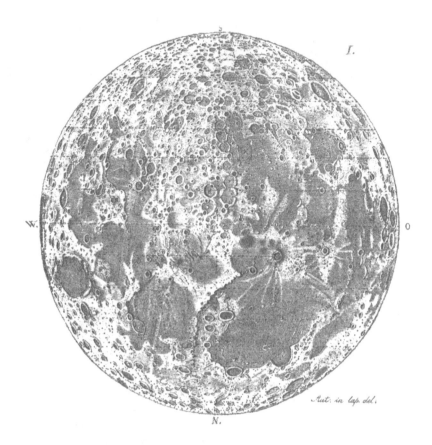

Fig. 73. Gruithuisen's version of Mayer's map, drawn and lithographed by himself (1825).

version was produced at some later date (fig. 74). Here we find a list of 144 names, all incomprehensibly joined to their appropriate features by a confusing network of straight lines! Of these names, 111 are from Riccioli, 12 from Hevelius, and 13 from Schröter. The remaining eight stem from Cassini (**Prom. Acutum**), Mayer (**Heraclides falsus** and **Mercurius falsus**), Lohrmann (**Bode**), and Gruithuisen himself (**Cometicus, Herschel, Schroeterus,** and **Sinus Medius**). **Cometicus** very aptly applies to the crater pair Messier and Messier A of our modern maps, plus the eye-catching 'comet-tail' streaks that stretch westward from those; **Herschel** is named Hedin nowadays; **Schroeterus** refers to the 'city' formation although, as we shall see, that name would soon be moved to a nearby crater; and **Sinus Medius** to modern Sinus Medii. Readers may recall that Van Langren used that form of the name in his map of 1645, but it is doubtful that Gruithuisen ever saw a copy of that map. Thus over the course of time, this small dark bay-like area at the centre of the Moon's disc has

Fig. 74. (turned opposite) A later and apparently very rare edition of his map with an almost impossible scheme for identifying feature names.

Zu Tab. I. gehörig

Register der Mondflecken

N

gone through the following series of name changes: **Insula Medilunaria**, **Sinus Medius**, **Mare Adriaticum**, **Sinus Aestuum**, **Insula Sinus Medii**, **Sinus Medii**, **Sinus Medius**, and finally back to **Sinus Medii** today. Hardly conducive to a stable nomenclature, but indicative of things to come by the close of the nineteenth century. According one source, Gruithuisen also applied the following names to features: **Cascade**, **Eisenhard**, **Keill**, and **Moenum**, but this needs to be checked.

CHAPTER 7

....

LUNAR CARTOGRAPHY COMES OF AGE

LOHRMANN SHOWS THE WAY

The observational work of Schröter and, to a lesser extent, that of Gruithuisen, demonstrated the large amount of surface detail that could be detected on the Moon with the ever-increasing quality of the telescopes of the day. Russell had made a valiant stab at producing scientifically accurate lunar maps and globe, but these did not pretend to include the mass of finer detail visible in his telescopes. Furthermore, his maps did not include lines of latitude and longitude, certainly a necessity for scientific purposes, and their great rarity would have limited such usage anyway.

No doubt the daunting problem of making both accurate position measurements of a large number of surface landmarks and calculating their absolute coordinates, coupled with portraying much more detail than that shown on the Tobias Mayer map, was enough to discourage all but the most dedicated persons.

The first to take on the challenging task of constructing a new, large scale map of the Moon, based on a larger and more accurate network of measured positions and including both finer and more accurately portrayed detail, was a Dresden surveyor/cartographer, Wilhelm Lohrmann (1796–1840). After making a map of part of the lunar Apennine Mts, using established cartographic practices on observations made with a 4.8-inch-diameter refractor in early 1822, he finally planned to construct a map in 25 square sections, laid out and numbered as in fig. 75, with a lunar diameter of 37.6 inches. The first four sections appeared in 1824, in his *Topographie der sichtbaren Mondoberflaeche*, which includes descriptions of the formations and features depicted. In fig. 76, which illustrates Section

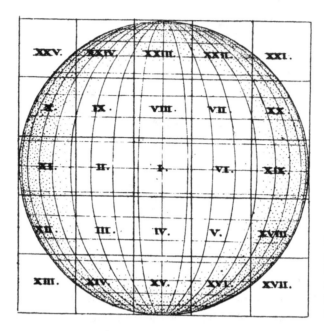

Fig. 75. Lohrman's scheme for dividing the Moon's disc into 25 squares for mapping at larger scale in manageable sheets.

III of this atlas, we see his method of portrayal of both highland and mare areas. Here we find, for the first time in lunar maps, the use of hachures to indicate the length and steepness of the slopes of the various features. This technique had been systematized and brought into common usage for terrestrial cartography by J.G. Lehmann (1722–1805) only 25 years earlier, but had the drawbacks of being extremely labour intensive and, ideally, requiring an artistic engraver who was sensitive to the telescopic appearance of the lunar surface (fig. 77a). Variations in surface reflectivity are indicated by carefully applied stippling. South is up, thus following Schröter's convention.

Unfortunately, Lohrmann's pioneering work was never completed during his lifetime. Most accounts attribute this to the onset of blindness, but more recently it has been noted that he more likely died from a typhoid epidemic. He had, however, completed drawings of all the 25 sections of the map four years before his death, and published a wall map at exactly 40 per cent of the scale of the main map during this period.

Noting that the cartographical part of Lohrmann's painstaking work was essentially complete in MS form, another German selenographer, J.F. Julius Schmidt (1825–84), took it upon himself to publish the remaining 21 sections of the map. This was no minor undertaking, in view of the tremendous amount of highly skilled effort needed to execute the engrav-

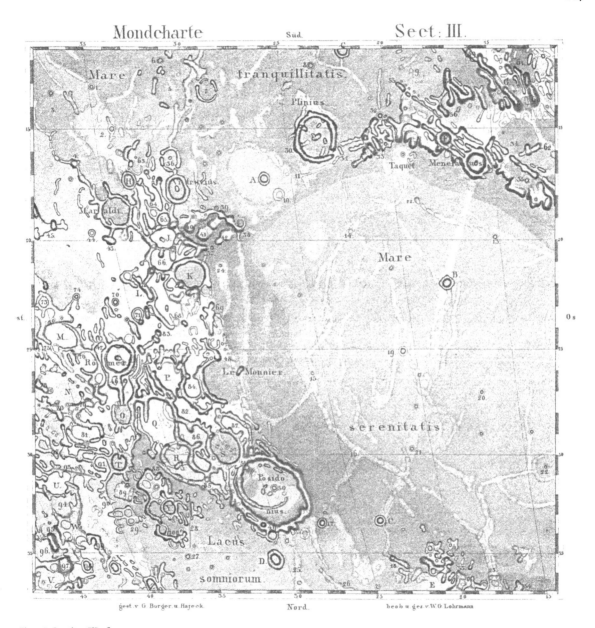

Fig. 76. Section III of
Lohrmann's map (1824),
showing names and many
numbers and letters.

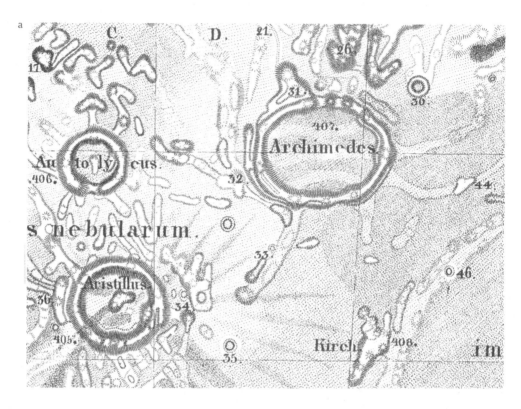

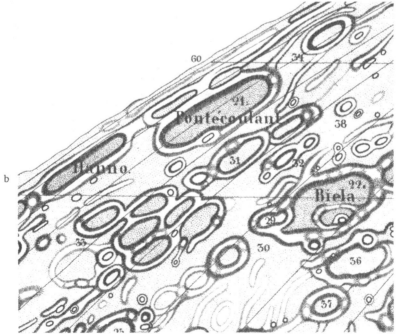

Fig. 77. Enlarged area of section IV of Lohrmann's map showing (a) the meticulous hachuring and shading, and (b) the lack of perspective effect in craters situated near the limb.

ings. The names of eight engravers and one advisor are recorded on the various sheets. The final product did not appear until 1878, by which time other maps and atlases had superseded it.

REVIEW OF THE LOHRMANN MAP

Examination of all the sections of this map shows that it is indeed a worthy successor to Mayer's pioneering map, which was drawn up some 75 years earlier. The hachuring technique works very well for the Moon, where relative altitudes depend entirely on the interpretation of shadows cast by the eminences and hollows. Comparison with modern photos (how easy we have it nowadays!) shows that the surface topography is remarkably accurately portrayed, except towards the limb, and misses very little of the detail that would have been visible with a telescope of 4.8 inches' aperture under average seeing conditions.

If the map falls short at all, it is in the portrayal of those craters in which the floor is notably lower than the surroundings, and also of areas nearer to the Moon's limb. In the first case, the impression imparted is of diametrically sliced doughnut halves (the type with holes) resting on a level surface. In the second, the sides of mountains and craters facing away from the Earth may be invisible to the observer, but they are portrayed in the map as being equally as wide as the sides facing the observer, thus losing the readily achievable perspective effect (fig. 77b). Lines of selenographical latitude and longitude are drawn at 5° intervals for each sheet, with 1° markings at the edges.

LOHRMANN'S NOMENCLATURE

It is hardly surprising that with the larger scale and improved resolution and accuracy of his maps, Lohrmann found it necessary to add some sort of lettering or numbering scheme to identify features for his verbal descriptions. Sections I–IV, the only ones that he saw published, follow Schröter by using the Riccioli names plus Hevelius's Montes Apenninus, but also contain eight new names, as follows: **Bode, Delambre, Dollond, Herschel, La Lande, Le Monnier, Maskelyne,** and **Triesnecker**. These have all remained unchanged to this day. Lesser formations were given capital Roman letters, from A to Z in the case of Section III (fig. 76), allotted in a semi-random order, with the minor features being numbered from 1 to 100 in a zig-zag pattern that makes it very difficult to locate any given number.

The wall map does not include any names other than the Riccioli 'watery' designations, but excludes Palus Nimborum. The main topographic formations are numbered from 1 at the south pole to 500 at the north pole. These numbers bear no relation to the numbers on the sections! In Schmidt's 1878 issue of the whole 25-section map, he has added these serial numbers to their appropriate formations. With Lohrmann's numbers and letters still intact from his completed manuscript sections, together with nearly all the names extant by that time, the confusing state of affairs is easy to imagine! Actually this was of little consequence, as yet another major selenographical work was begun in Germany in about 1828, when it became apparent that Lohrmann would be unable to complete his project. This one was destined to become a prime reference work for some 50 years.

BEER AND MÄDLER'S MAGNUM OPUS

As frequently happens in scientific enquiry, the advent of improved or updated techniques is a strong incentive to improve upon earlier researches that were made with more primitive instrumentation and were hampered by insufficient or just poor data. Sometimes two groups or persons will independently initiate a similar project, resulting in some degree of competition once they learn of each others intentions or work in progress. We have already seen the case of Van Langren, Fontana, and Hevelius, who were making observations of the Moon and preparing maps almost simultaneously. Hevelius, who probably had the best skills and knowledge, and certainly had by far the best resources at his disposal, eclipsed his contemporaries with his massive and widely available *Selenographia*.

Here we have rather similar circumstances. The young (26 years old) and enthusiastic cartographer Lohrmann started off in great fashion by completing and publishing (at his own cost, be it noted, as was the case for Schröter) the first four sections of his selenographical work in just two years. But with no further sections appearing during the following four years, another young person, this time an astronomer, Johann Heinrich Mädler (1794–1874), who apparently was at least as enthusiastic as Lohrmann, decided to attack the subject as thoroughly and expeditiously as possible. With the co-operation of his friend Wilhelm Beer (1797–1850), who owned his own observatory equipped with a 3.75 inch refractor, he completed a truly monumental achievement in the space of only nine years (1828–37).

A MAP AND A TOME

As a starting point, Mädler decided to produce a map at the same scale as that used by Lohrmann – about 37 inches – and to use the same technique of hachuring to depict the surface topography. Rather than dividing the map into 25 page-sized sections, however, he opted to produce it in four parts, each a quadrant of the Moon's disc, bounded by the lunar equator and prime meridian with no overlap. The lithographic process would be used for producing the actual copies of the map rather than engraved copperplates as used for the Lohrmann sections. His MS map was actually prepared to 104 separate sections, using the same hachuring scheme to be used in the final map. Readers might be interested to know that the existing 99 MS sheets were offered for sale in 1977 by a Liechtenstein bookseller for 62 000 Swiss francs, about £20 000! Quadrant 1 of the final map, the first to be published (in 1834) is shown at a much reduced scale in fig. 78. The remaining three quadrants were published through the following two years, the whole map (south up) being titled *Mappa Selenographica*.

This map, although greatly surpassing in detail and accuracy all of its predecessors, was only a part of the overall product of the nine years of intensive observation and research. The remainder was published in 1837 in a major book, *Der Mond [The Moon]*, which presents us with over 400 large pages, in comparatively small print, of just about everything that was known about the Moon at that time, including its complicated orbital motion, the librations, eclipses, etc., and the mathematics involved; full details on the measurement of positions of lunar features, of their heights, depths, diameters; an account of the history of observations of the Moon; the effect of the Moon on the tides and weather; evidence for or against a lunar atmosphere; the origin of the surface features and their physical nature; comprehensive notes and tables on the nomenclature; etc., etc. This material occupies about one half of the text, the remainder consisting of detailed descriptions of the features and formations of every area of the Moon's nearside. It is truly a monumental work, done with typical German thoroughness.

REVIEW OF THE MÄDLER MAP

Although the hachuring technique for feature portrayal is again used, it is more delicate than that found in the Lohrmann maps, resulting perhaps in a more natural appearance. A small area of the lunar surface as depicted by Lohrmann and Mädler, together with a photo of the same area for comparison, is illustrated in fig. 79a, b and c. The mare areas and other darker

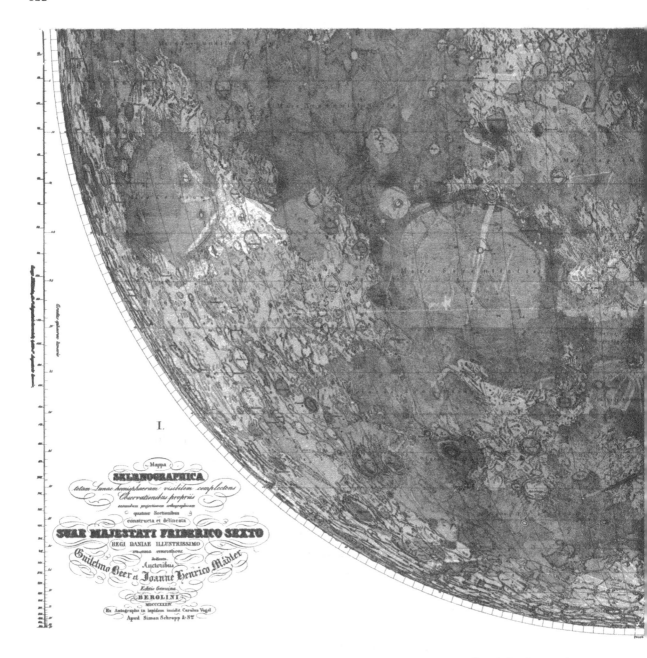

Fig. 78. Quadrant 1 of
Beer and Mädler's map
(1834), greatly reduced.

spots are stippled a little more densely than the highland areas, as in the Lohrmann map.

The map was re-issued in 1877, but not before some restoration had to be performed on the lithograph stones. The imagery on several areas mostly near the centre of the Moon's disc and along the inner edges, i.e. the equator and prime meridian, were apparently lost during storage, also on a thin strip 7 inches long near the lower left corner of quadrant 4. These missing areas were carefully drawn in once again, but comparison between the two editions shows a few interesting differences. Thus several craters show more detail in the later edition than in the first one, while some minor details were lost. The main problem, however, was that letters designating many small surface features were never replaced in the later edition. This introduced a number of nomenclatural problems some 30 or so years later, when three major lunar maps were being compared.

MÄDLER'S NOMENCLATURE

Mädler was confronted with much the same problem that Schröter and Lohrmann had to deal with earlier, namely, how to name or designate the large number of newly mapped formations, and to what extent the clumsy but occasionally useful Hevelius names should be used. In the event, he basically followed Schröter's lead by using Riccioli's nomenclature in preference to that of Hevelius, although he quotes and tabulates the Hevelius equivalents where possible throughout the text. Of the 426 names used, four are from Van Langren (although his map was never seen by Mädler), 202 from Riccioli, 10 from Hevelius, 68 from Schröter, one from Gruithuisen, eight from Lohrmann; the remaining 133 are new names added by Mädler 'which are taken from astronomers, geographers, mathematicians, and naturalists of recent and partly also of earlier date.' One of these names is none other than **Flamsteed**, which we have encountered before in the obscure maps of Keill and Hell. Mädler's new names are listed in Appendix I.

Mädler broke with tradition in changing the latinized form of a few names from Riccioli back into their original forms (e.g. Hevel, Kepler, rather than Hevelius, Keplerus). He also converted some names into a form that was best suited to German users, so we notice such spellings as Ariadäus, Iulius Cäsar, Fracastor, Ptolemäus, Galiläi, etc. This last case is particularly interesting, because it illustrates two undesirable changes. Galileo's surname in Italian was Galilei; when latinized, this became Galileus or more frequently, Galilaeus. The gentive form ('of Galilei') is

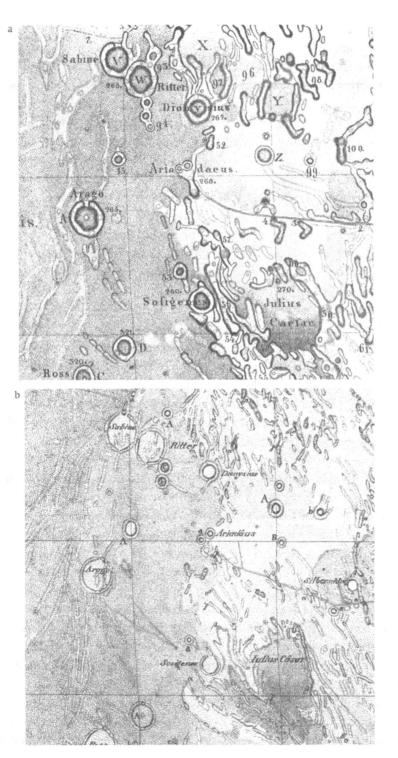

Fig. 79. Enlargements of an area common to Lohrmann's (a) and Mädler's (b) maps showing the difference in style and interpretation, with a photograph for comparison (c). The same area as depicted by later observers is given in fig. 87.

Fig. 79. (*cont.*)

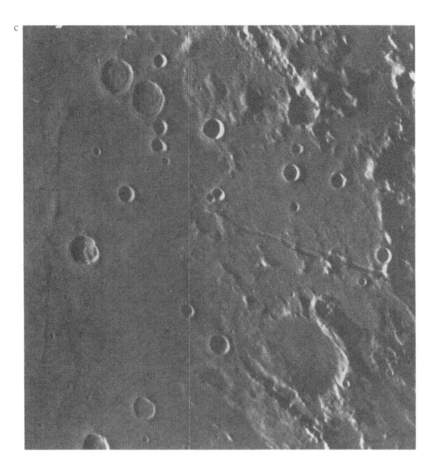

'Galilaei'. Mädler put this in German form with umlaut, 'Galiläi', which reverted to the Latin genitive when later anglicized. The other change is that Riccioli applied the name to the prominent bright 'light-bulb' marking nowadays designated as 'Reiner Gamma'. As Mädler did not give names to albedo markings he moved the name 'Galiläi' to the nearest crater, where it has remained to this day. Thus one of the greatest astronomical pioneers is commemorated by a miserable little crater, with incorrect spelling to boot! Just to complete the confusion, our Reiner Gamma is not even the ridge originally so designated by Mädler! But we are getting ahead of ourselves.

Named features on the Riccioli and Schröter maps which Mädler could not identify, or which he thought were not worthy of having separate names, such as areas in the brighter highlands regions, were divested of those names. A few of those names, however, were transferred to other

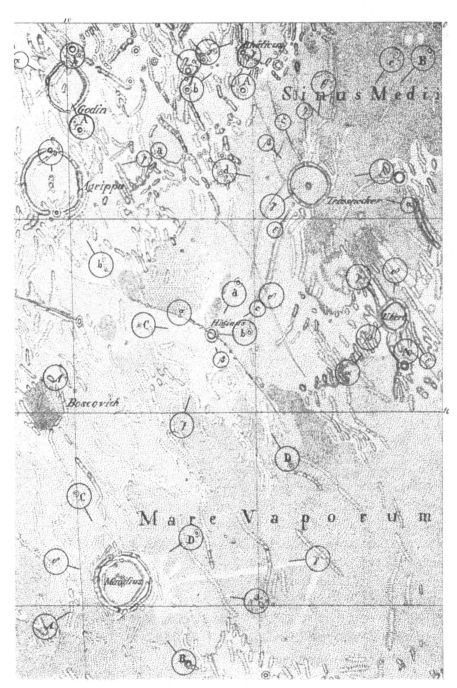

Fig. 80. Another enlarged area from Mädler's map illustrating his method of lettering subsidiary features by placing the letters on the side towards the 'patronymic' crater. The circles and directional lines have been added for clarity.

features. By inter-comparing the Riccioli, Schröter, and Mädler maps with modern photographs taken at various phases, it is readily seen that several mis-identifications were made by both of the later observers. As I have noted before, it is easy to criticize when these comparisons can be made indoors in comfort. Schröter's and Lohrmann's specific numbers and letters were not used, partly because of their total inconsistency, but also 'as we made use of them' [i.e. the numbers] 'for the lines of latitude and longitude and, there was frequently no space for them'.

Mädler's scheme for designating subsidiary features differs from Schröter's in that he uses qualified letters, i.e. all letters are preceded by the name of the nearest named feature when used in texts, catalogues, etc., or orally. The method of applying letters to the map is noteworthy, since it both obviates repetition of the names of the 'patronymic' feature, thereby minimizing obscuration of surface detail, and also provides information regarding the nature and placement of the feature.

In this scheme, subsidiary craters are given Roman letters, capitals for those whose selenographical coordinates had been determined, lower case letters for those whose map locations were eye estimates only. Subsidiary mountain peaks or blocks, points on crater rims with measured heights, mare ridges and rilles are given Greek letters, again with capitals for those points with measured coordinates (fig. 80). Although Mädler states that the letters are allocated in approximate order of importance for each group of subsidiary features, the angle of solar illumination has such a profound effect on the relative visibility of features that his statement is almost meaningless. Thus a less desirable feature of the system is the apparently haphazard disposition of letters, with no obvious regard to the relative location or importance of the subsidiary features, and with gaps in the alphabetical sequences in about 9 per cent of the cases. However, similar criticisms may be levelled at the Bayer letters for the constellations. There are 3000 lettered features in all, 1551 being craters, 1449 hills, rilles etc.

THIRD ERA

FROM PROLIFERATION TO
STANDARDIZATION

CHAPTER 8

LUNAR MAPPING IN THE
VICTORIAN PERIOD

SCHMIDT DOUBLES THE SCALE . . .

The first reaction to Beer and Mädler's monumental work was one of resignation – everything that could be learned about the Moon was there for anyone to read, and everything that could be observed with a telescope of moderate size was already mapped. Their pronouncement that the Moon's surface was nothing but a dead, volcanic wasteland stifled further physical investigations of the nature of that surface for a while, steering subsequent observation and research into checking and improving the map and descriptions of the features. One of the first to proceed along these lines was J.F. Julius Schmidt who, as we have seen already, involved himself in the publication of Lohrmann's manuscript atlas sheets. Beginning in about 1840, he observed the Moon assiduously for some 34 years, making numerous drawings (fig. 81) and height and depth measurements.

All this intensive work culminated in 1878 with the publication of a map and book, *Charte der Gebirge des Mondes*. The format of the map follows that initiated by Lohrmann, i.e. 25 square sections with no overlap, south-up, with topography indicated by hachuring. The map areas are coloured with a pleasant sepia tone, with the darker patches emphasized either by stippling or by deeper sepia ink. Degree divisions are marked at the edges of the map, but the 5° latitude and longitude lines are not drawn across the map as in the case of Lohrmann. Schmidt used the position measures previously made by both Lohrmann and Mädler, which really were insufficient for a map at the scale that he chose, namely about twice that of Lohrmann's and Mädler's maps (actually 77.5 inches). Section III is illustrated at greatly reduced scale in fig. 82.

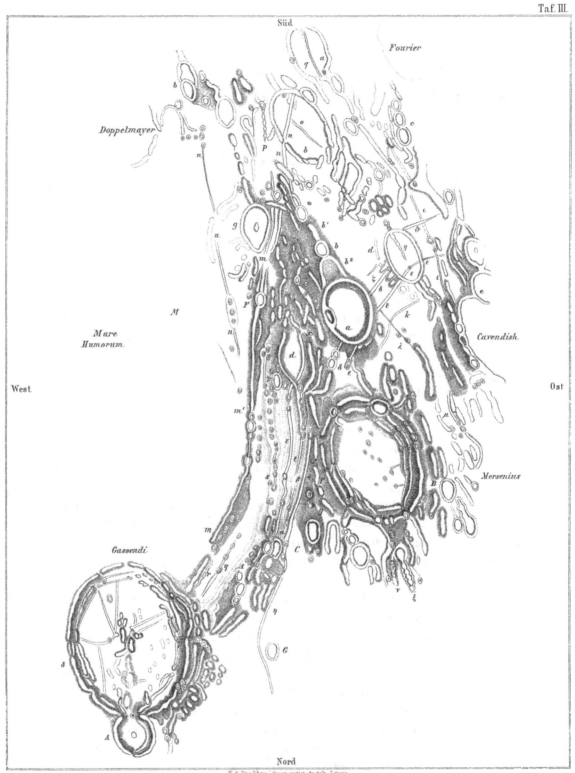

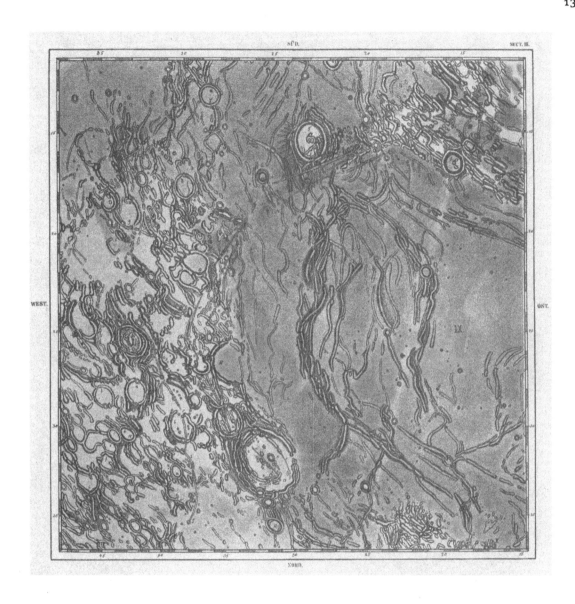

Fig. 81. (opposite) A beautifully hachured engraving from a drawing by Schmidt.

Fig. 82. (above) Section III of Schmidt's great map (1878). Compare this with Lohrmann's map, fig. 76.

Schmidt employed telescopes that were generally of greater aperture than those used by Lohrmann and Mädler, allowing him to observe smaller details. This is evident when comparing the three maps, Schmidt recording numerous small features that are absent from the other two. Indeed, he notes that while Lohrmann shows 7178 craters and Mädler 7735, he depicts 32 856! The chief criticism that may be levelled at the map is that the low mare ridges are depicted as being as prominent as the much bolder crater rims. Nonetheless, this map is undoubtedly the finest that

Fig. 83. A portion of
Schmidt's map showing
the less formal hachuring
and barely discernible
lettering (circled).

was ever compiled without the aid of photographs. The text of the accompanying book is quite unlike Mädler's, consisting almost entirely of height measurements, with a little space given to nomenclature comparisons and general notes. The map was accompanied by a booklet titled *Kurze Erlauterung zu J. Schmidt's Mondcharte*, which lists the named features for each of the 25 sections, and gives brief descriptions of the areas covered.

SCHMIDT'S NOMENCLATURE

Not wishing to cover up useful topographic details in his map, Schmidt opted to use numbers for the named craters on the map, the key being given in both the book and booklet. He used the Mädler nomenclature, including the letters, but added 76 new names, many to features near the limb that were poorly depicted or absent from the Mädler map, or to lettered features on that map. These are listed in Appendix L. He did not follow Mädler's lettering meticulously, omitting some and adding new ones. They are quite unobtrusive on the map, and are easily missed (fig. 83). He reverted to using the diphthong 'ae' rather than Mädler's somewhat contrived 'a' with umlaut in the case of the older latinized names.

Things did not turn out to be quite straightforward for Schmidt because two other selenographical projects were under way in England towards the later stages of the production of his map. These were, first, an ambitious programme conceived by the British Association (BA) to catalogue and map the lunar surface in the finest detail possible, and second, an unrelated effort, by E. Neison, to produce an abridged translation of Beer and Mädler's *Der Mond*, with improvements and updates plus a simplified map in 22 page-sized sections. Before seeing how these two projects impacted Schmidt's nomenclature, and in fact each other's, let us look at them a little closer.

. . . AND THE BA QUINTUPLES IT

The British Association resolution to map the Moon in great detail dates to 1865, but the whole project was doomed from the very start because of the unbelievably ponderous scheme proposed to draw up the map, and to designate the various features. For the original drawings, a scale of 200 inches to the Moon's diameter was chosen. I cannot resist quoting the description of this scheme given by E. Neison, who is the author of the second project noted above.

> On the British Association system the division into four quadrants by Beer an Mädler is retained, they being numbered from I to IV in the following

order: NW, NE, SE and SW. Each of these quadrants is divided into 16 grand divisions, distinguished by the capital letters to A to Q, and consisting of an area 25° square, except towards the limb, where of necessity only 15° remain on this hemisphere, and the remaining 10° extend into the further side, which is brought into view by libration. The lettering runs from the equator to the poles, so that B stands nearer the pole than A; but between the same circles of longitude, whilst the square, on the side of A, between the equator and the 25° parallel of latitude, only nearer the limb, is E. Each of these grand divisions of 25° is further sub divided into 25 areas of 5° square, lettered in the same manner as the grand divisions with the Greek letters alpha to omega, the last space being left blank. Finally, any object is distinguished by a number attached to the symbol denoting the small area of 5° square which it is in, and also the quadrant. Thus IAσ40 would indicate object No.40 in area IAσ; that is, between the limits of 10° to 15° west longitude, and 15° to 20° north latitude.

Obfuscatory verbiage worthy of a legal document!

Work on this project was undertaken by William R. Birt (1804–81), a meteorologist and amateur selenographer. Very few of the 5° squares were completed, and the whole project fell through. One of these sections is reproduced in fig. 84, together with a photo of the same area for comparison. The lines on the map will be seen to be almost totally meaningless, another reason for its failure. However, Birt did have quite an influence on nomenclature outside this project. He drew a rough outline map of the Moon, based loosely on the Mädler map with their nomenclature (fig. 90), adding, together with his friend John Lee (1783–1866), 85 new names (Appendix J). It was soon discovered that several of these had also been used by Schmidt, but that they had been placed on different features. Similarly, a given formation sometimes received different names on the two maps. But there was more trouble coming.

NEISON'S POPULAR BOOK

The second project mentioned above, a volume titled *The Moon*, by Edmund Neison (1851–1938), was being prepared concurrently with Schmidt's and Birt's, and was published in 1876, just two years before Schmidt's. Its main purpose was to provide a paraphrased translation into English of the descriptive material of Beer and Mädler's text, but with improvements and additions based on Neison's own observations and also some of those by Schmidt, and with map sections that were the size of the book pages. This was accomplished by dividing the Moon's disc into 22 rectangular sections (fig. 85) and choosing a scale of 24 inches to the lunar diameter. At

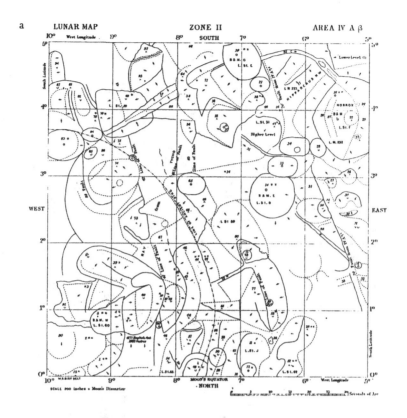

Fig. 84. One section of the never finished British Association's map as drawn up by Birt (a), with a photo of the same area for comparison (b). The lines, arrows, letters etc. will be seen to be almost completely meaningless.

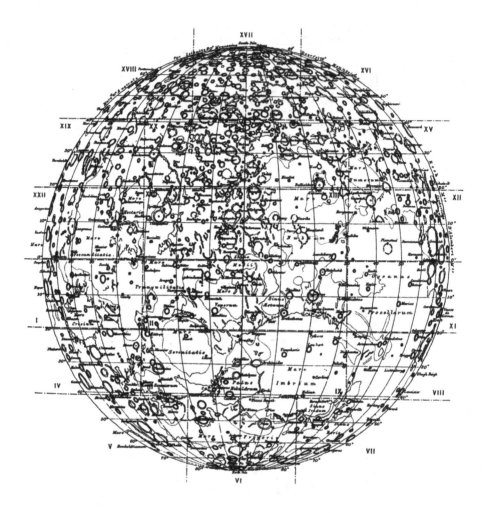

Fig. 85. Neison's scheme for dividing the lunar disk into rectangular sections to correspond with the shape of his book pages.

two-thirds the scale used by Mädler and with more fine detail to record, Neison was obliged to come to some sort of compromise. He did this by omitting all albedo differences, simplifying the more nondescript areas, and depicting the newly observed finer detail at a scale somewhat larger than reality (fig. 86). He chose the hachuring method to depict the surface topography, mimicking Mädler's style rather than that of Schmidt or Lohrmann, but rilles are shown as parallel lines, usually linear or with minimal bends. Figure 87 shows (a) Neison's and (b) Schmidt's versions the same area as that shown in fig. 79, as depicted on Mädler's, Schmidt's, Neison's maps.

Neison added 15 new names to his map (Appendix K), but also reinstated a few of Riccioli's and Schröter's. With three projects under way at the

Map XIII.

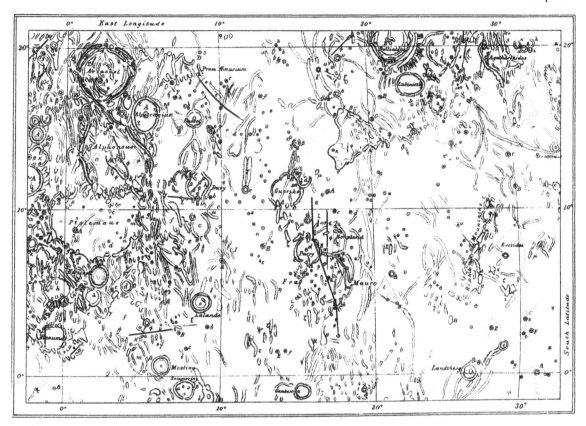

Fig. 86. Section XIII of Neison's map.

same time, and each adding new names, it is not hard to understand the nomenclatural chaos that was beginning to ensue. There was some exchange of documents between the parties involved, since Schmidt and Neison each acknowledge a few of the additions of the other two, but this was understandably incomplete because of the concurrency of the projects.

PHOTOGRAPHY ENTERS THE PICTURE

Early attempts at photographing the Moon through telescopes were particularly difficult, because of the insensitivity of the collodion emulsions at that time, entailing comparatively long exposures that meant guiding the telescope at the lunar rate, and being at the mercy of the blurring effects of atmospheric 'seeing'. Figure 88 shows some of these images, dating from about 1851 to 1873. Not one was in serious contention with

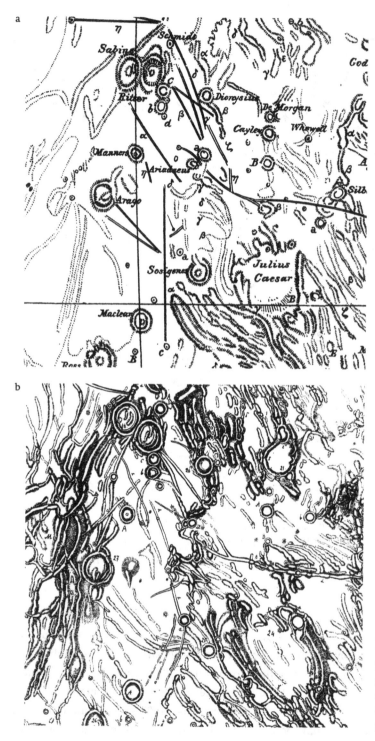

Fig. 87. Same area of the lunar surface as shown in fig. 79. Neison's (1876) depiction, with straight 'tramline' rilles and a liberal addition of letters, is shown considerably enlarged in (a), while Schmidt's version is given in (b).

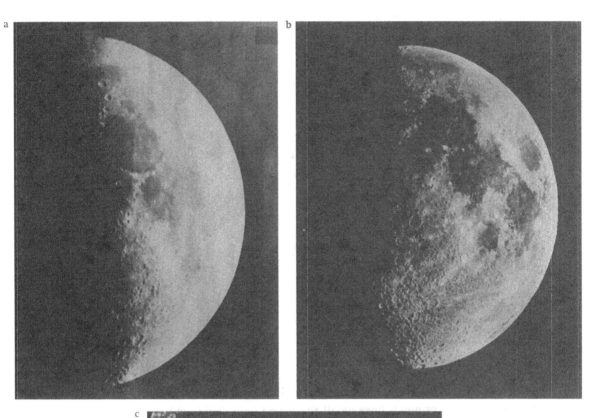

Fig. 88. (this page and overleaf) Copies of some early photographs of the Moon. (a) and (b) are by Bond at Harvard University, and date from the early 1850s; (c) is by De la Rue, Oxford, c. 1875 (d) is by Rutherfurd, New York, 1870; (e) was taken in 1873 with the ill-fated and short-lived Great Melbourne reflector.

visual observation, the eye exerting a degree of filtering of the seeing and thus registering very much finer detail than the photo could achieve. However, the photos had the distinct advantage of being permanent and stable records that could be examined and measured at leisure indoors. By the close of the nineteenth century, photographic emulsions had improved to the degree that position measurements of lunar features made on them superseded both in number and accuracy the old eye-at-the-telescope measurements of Lohrmann, Mädler, and Neison. But the photos still did not show details smaller than could be seen visually with quite modest sized telescopes, say 2 to 3 inches at most, and thus smaller than the instruments used by those selenographers.

Fig. 88. (*cont.*)

A LARGE GLOBE, AND PLASTER MODELS

Before entering on twentieth century progress in lunar observational studies, we need to take a brief look at a few other projects that date from about 1840 to 1900. Mention should first be made of the fine general chart issued by Mädler in 1837 (fig. 89), with a scale of one-third that of the large map, about 12.5 inches diameter. This was used as a basis for several grossly inferior outline maps, such as that by Birt (fig. 90).

Fig. 89. (opposite) Mädler's excellent smaller map (1837) with 368 numbered features, and many measured heights and depths.

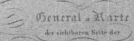

General-Karte
der sichtbaren Seite der
MONDOBERFLAECHE
zugleich als
ÜBERSICHTS-BLATT
zur grössern Mondkarte von Wilh. Beer und Joh. Heinr. Mädler,
gezeichnet von J. H. Mädler.
BERLIN 1837

Erläuterungen.

Berlin, bei Simon Schropp et Comp.

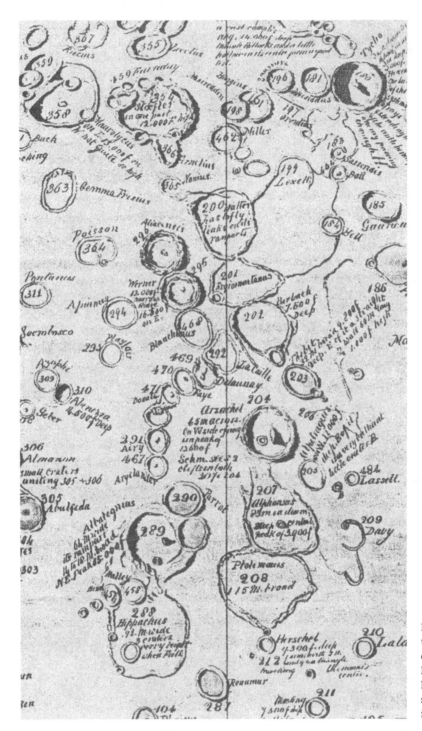

Fig. 90. A portion of Birt's unpublished, poorly drawn map of the whole lunar disc, with his handwritten remarks sprinkled between and inside craters.

Fig. 91. The 19-ft-diameter globe of the Moon made in Bonn by Dickert in 1850. The plaster surface was modelled after Mädler's map, but the surface markings shown here display all the idiosyncrasies of Russell's full-Moon image!

Relief der sichtbaren Halbkugel des Mondes, angefertigt von Th. Dickert in Bonn.

A little-known hemispherical model of the Moon's nearside was produced in Bonn, Germany, in the early 1850s by Th. Dickert, with supervision and additions by Schmidt. It was 19 ft in diameter, and was based on Beer and Mädler's large map. The plaster surface was coloured matte yellow, and the three-dimensional relief was amplified threefold. Figure 91 is an illustration of this globe from a contemporary magazine (*Illustrirte Zeitung*, No. 589, 14 October 1854). It is interesting to note that the markings shown in this image show the idiosyncrasies of Russell's image of the full Moon (see fig. 63)! One wonders if that image was used on the globe, or just for the illustration; certainly the albedo shadings on Mädler's map are very muted, and not conducive to a natural appearance for an exhibited

Fig. 92. Photograph of a plaster model of the lunar Apennines and adjacent region made by James Nasmyth in about 1872.

model. The globe was transported to America, and was last known to be at the Field Columbian Museum in Chicago.

Another important lunar study from this period is a book devoted to the nature and possible origin of the surface features. This was the work of James Nasmyth (1808–90), better known as the inventor of the steam hammer and a convenient form of altazimuth telescope mounting, and James Carpenter (1840–99). Their book, *The Moon*, was published in 1874, and is notable, especially in later editions, for its tipped-in photos of excellent plaster models of several areas of the lunar surface (fig. 92). The authors espouse the volcanic theory for the origin of the craters, the crater

Map of the Moon
by
T. GWYN ELGER, F.R.A.S.

FIRST QUADRANT.

GEORGE PHILIP & SON LONDON & LIVERPOOL.

Fig. 93. Quadrant 1 of Elger's popular Moon map.

walls resulting from the piling up of ejecta from the central mountain, which supposedly had sprayed out in the form of a hollow cone.

Sensing the need for a smaller, popular guide to the Moon than Neison's bulky volume, T.G. Elger (1838–97) produced such a book in 1895 (*The Moon*). It contains an artistic, very clear, map in four quadrants, 18 inches in diameter (fig. 93). It was also available as a single separate sheet. This publication proved to be very popular with observers, the map being reprinted in the 1950s following the addition of some poorly executed extra details.

As mentioned earlier, photography of the Moon had improved by the

Fig. 94. (opposite) Moon map by Lecouturier and Chapuis (1860). Not a particularly accurate or aesthetically pleasing effort, but the name 'Leverrier' was introduced for the first time on this map.

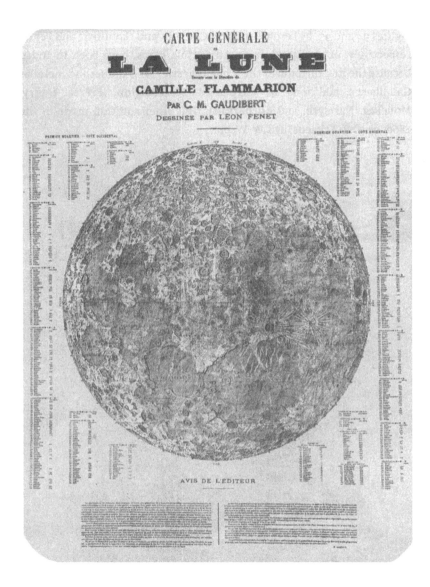

Fig. 95. Another French map from about 1890, this one including six new names.

close of the nineteenth century to the point where it constituted not only a useful supplement to visual observations, but also provided a means of making accurate position measurements of a much larger network of points on which to base newer and better maps. Various new lunar studies had got under way or were completed by this time, but remembering that in this book, we are concentrating on the progress (and regress) of the nomenclature and related cartography, we must reluctantly give little or no attention to many worthy projects that do not meet these criteria.

Figures 94 and 95, by Lecouturier and Chapuis, and Gaudibert and Fenet, respectively, otherwise quite unremarkable maps, have been included because the first introduces the name **Leverrier**, and the second includes Gaudibert's additions of **Carpenter**, **Flammarion**, **Frères Henry**, **Mouchez**, **Nasmyth**, and **Rutherfurd**. One other addition made in the 1860s was **Mt. Argaeus**, by T.W. Webb.

CHAPTER 9

NOMENCLATURE GETS INTERNATIONAL
ATTENTION

FRANZ AND SAUNDER LEAD THE WAY

The last decade of the nineteenth century saw two programmes of lunar photography that were intended to provide coverage of the lunar surface at various phases for pictorial and research purposes. The smaller programmes was carried out at Lick Observatory, culminating with a small atlas of mediocre quality enlargements. The major programme, however, was undertaken at Paris Observatory by M. Loewy (1833–1907) and P. Puiseux (1855–1928), who designed and used a novel telescope (the 24-inch equatorial coudé refractor) to take some 6000 plates of the Moon. The best were used to produce the famous but rather unwieldy *Atlas Photographique de la Lune*, consisting of 80 enlargements by the heliogravure process on sheets almost 2 × 3 ft in size. The degree of enlargement was such that the individual grains of the negative are reproduced!

Julius Franz (1847–1913), a professional astronomer at Breslau Observatory, acquired original plates from Lick, and used them to make position measurements of many formations and landmarks mostly towards the limb of the Moon. Some of the plates taken under conditions of strong lunar libration displayed features not drawn on any maps, enabling him to measure and map them. Many were patches of darker mare material, which, being anonymous, were given names by him, many of which were quite whimsical or descriptive in nature. These, together with a few other additions from this late-1800s era, are tabulated in Appendix M. He also gave letters to all the smaller craters and spots that he measured. In his final catalogue of coordinates, which are given as selenographic latitudes and longitudes, he includes four quadrant outline

maps, each in an orthographic projection constructed for favourable librations of 10° in latitude and longitude. Figure 96 illustrates quadrant 1.

Roughly contemporaneous with Franz was Samuel A. Saunder (1852–1912), a college-level teacher of mathematics and an amateur selenographer, who firstly obtained plates from the Paris collection, and shortly afterwards, two plates with superior resolution taken with the 40-inch refractor at Yerkes Observatory. He used all of these to make measure-

Fig. 96. (opposite)
Quadrant 1 of Franz's
map, based on his meas-
urement of numerous
lunar features, especially
those lying near the limb,
from several original
negatives taken at Lick
Observatory. Note that
the map is 'librated' 10°
in both latitude and
longitude so that the
limb areas are opened
out more.

ments of over 3000 points, mostly more towards the central regions, thus complementing those of Franz. It was during this programme that he encountered the duplications and other problems of the nomenclature, not only those arising from differences between Mädler, Schmidt, Neison, and Birt, but now also those from Franz.

Here is an example of this confusion: Schmidt gave the name 'Beer' to Mädler's 'Fracastorius E', which Neison named 'Rosse'; however, Neison gave the name 'Beer' to one member of Mädler's crater pair 'Archimedes B', which Schmidt named 'Hamilton'. Birt's 'Rosse' is a nondescript smooth area south of Phocylides. To complete this confusion, the other member of Mädler's close pair of craters that he had labelled 'Archimedes B' was named 'Mädler' by Birt, 'Beer A' by Neison, and 'Feuillé' by Schmidt! There are many other similar cases that can readily be found.

SAUNDER PROVIDES THE IMPETUS

Saunder brought this situation to the attention of the Royal Astronomical Society (RAS) in 1905 in section 11 of his first catalogue of positions. Another paper followed at the end of that year, and a year after that he wrote a third paper, for presentation to the General Assembly of the International Association of Academies, which was due to convene in Vienna, May 1907.

As a direct result of Saunder's notices, which also had the backing of the Royal Society, an eight-member committee, with international represen-tation, was set up to examine the whole question of lunar nomenclature. Its secretary and, from 1910, chairman, was Herbert H. Turner, Director of Oxford University Observatory. The main argument, given in the form of four questions, boiled down to this:

> shall an essentially new system of nomenclature be propounded, or do astronomers prefer to retain simply the traditional system, with necessary corrections and possible modifications?

Some of the ideas submitted in reply to this hark back to the unattrac-tive system tried by Birt some 40 years earlier. Thus Franz suggested that longitude sectors in 10° increments be given consonants, while lat-itude segments receive double vowels. With this system, the crater 'Short' becomes 'anal' (some censorship might be appropriate in some cases), and the crater 'Stevinus' comes out as 'edam' – perhaps lending a little support to the ancient green cheese concept! His scheme, like

Birt's, is a model of impractical complication. In similar vein, Puiseux and W.H. Pickering (1858–1938) also suggested alphanumeric systems. Saunder, however, proposed making no radical changes, using the basic Mädler names but retaining generally accepted new names, and using capital Roman letters for 'craters, depressions, and dark areas', and small Greek letters for 'peaks and bright spots'. He also pointed out the need for a new map to record the nomenclature once it had been agreed upon.

The same four questions were presented to the astronomical community at large through the *Monthly Notices* of the RAS in 1908. The general suggestions put forward by Saunder were overwhelmingly favoured, including the production of a map in mean libration. Franz and Saunder were asked to co-operate on the production of this, the former dealing with the outer regions and Saunder the inner. Saunder set down all of his fundamental points, and three of the four central squares had been artistically drawn in ink by W.H. Wesley, an able draughtsman, by 1913 (fig. 97). Note that here we do not see coordinates in the traditional latitude/longitude system, but rather in the familiar rectangular x/y system of ordinary graphs. This was introduced by Turner to simplify (a), the mathematics of converting x/y measurements made on the plates into the corresponding x and y coordinates on a Moon map oriented for mean libration; and (b), the plotting of those measured points on the paper. The coordinates are referred to as xi and eta; in this system, the third coordinate (zeta), perpendicular to xi and eta, is easily calculated if the lunar surface is assumed to be spherical. With good-quality photographs taken under different librations, this can give determinations of deviations from sphericity of the Moon's near side.

Franz completed his position measures, but drew his own map in four favourably librated quadrants as already noted (fig. 96). These, plus his position catalogue, were published in 1913 under the title *Die Randlandschaften des Mondes* [*The Border Regions of the Moon*].

MARY BLAGG DOES THE DONKEY WORK

For the major task of cataloguing and intercomparing the maps of Mädler, Schmidt, and Neison, plus taking account of Birt's many new names and any others that had crept into the post-Mädler literature, Saunder secured the services of Mary A. Blagg (1858–1944), who completed this daunting task also in 1913. She apparently had a copy of Beer and Mädler's *Der Mond*, but only the 1877 version of their map which lacked many letters and

Fig. 97. (opposite) A portion of the original map drawn by Wesley as part of the projected International Astronomical Union's map. It is based on the accurate measures of Saunder and Franz, and plotted in the rectangular xi/eta system, the squares representing one-tenth of the lunar radius.

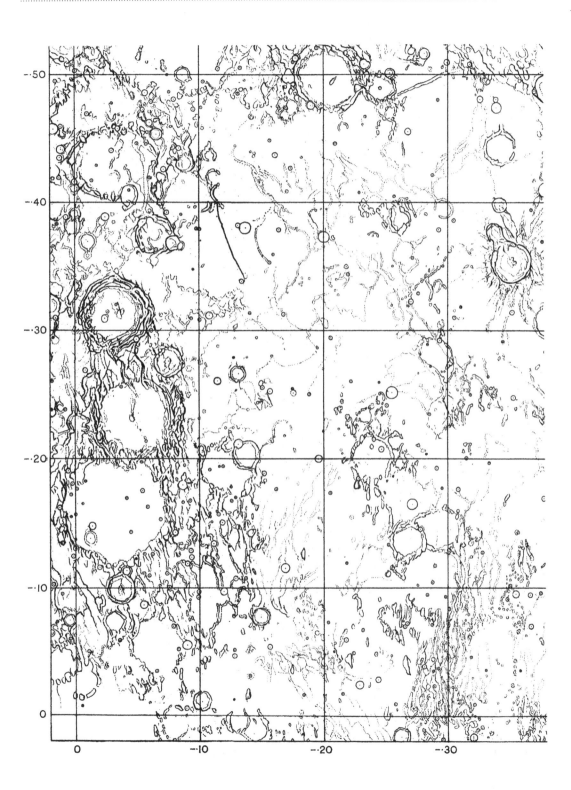

other details in certain areas. She also had Schmidt's text and map, Neison's book, but apparently nothing better than 'the Belgian Atlas', a much reduced and poor quality half-tone version of the Paris Atlas, for consultation.

Despite these handicaps, and an occasional blind eye when it came to spelling, her *Collated List of Lunar Formations* (1913) represents the fruits of an indefatigable enthusiasm for what must have seemed to be, with its 4789 entries (entailing over 14 000 comparisons between the three maps, plus uncounted references to the relevant texts and available photographs), an almost interminable chore (fig. 98). Wesley drew the fourth central quadrant at about this time, but the onset of the First World War in 1914 together with the deaths of Saunder, Franz, and other members of the Committee, brought all further progress to an abrupt halt.

GOODACRE GOES CARTESIAN

Before dealing with the important events that followed the First World War, we need to backtrack a little to 1910, when amateur astronomer Walter Goodacre (1856–1938) published a new lunar map. This has the distinction of being the first map to be based on accurate position measurements made from photographs, using the new xi/eta coordinates from Saunder's first catalogue of positions plotted on a rectangular grid. He chose a scale of 77 inches to the lunar diameter, the same as that employed by Schmidt, dividing the disc into 25 squares just as Lohrmann and Schmidt had done. However, in the published version the diameter was reduced to 60 inches, resulting in sheets about 18 inches square. Grid lines are drawn at intervals of 0.1 of the lunar radius in both xi and eta, and all sections have overlaps with adjacent sections.

Rather than using hachuring to depict surface topography, which is labour intensive and requires considerable artistic skill, he chose the form-line approach in which changes of slope are indicated by full or dotted lines, fig. 99. As with Neison's map, no albedo markings are included. Nomenclature follows that of Neison, but Goodacre added a number of capital letters to very low or ruined formations and enclosures that were previously anonymous.

THE INTERNATIONAL ASTRONOMICAL UNION TAKES OVER

In 1919, shortly after the end of the First World War, the International Astronomical Union (IAU) was founded to coordinate astronomical research and observation worldwide. Thirtytwo Commissions were insti-

NEISON'S, SCHMIDT'S, AND MÄDLER'S LUNAR MAPS. 73

	Neison.		Schmidt.		B. and Mädler.		Position.	Symbol.
2021	Rocca	327	Rocca (XIX.-15)	...	Rocca	359	...	o
2022	,, a	327	Rocca a (not named in map)	30	,, a	359	S.W. of Roc.	o
2023	,, B	327	(Not named)	...	,, B	359	W. of Roc.	o
2024	,, c or C	235–7	,,	...	,, C	359	N.W. of Roc.	o
2025	,, d	325	,,	...	,, d	...	W. of 2024	o
2026	,, e	327	,,	...	(Not named)	...	N.E. of 2023	o
2027	,, f	328	,,	...	Rocca f	359	W. of 2022	o
2028	,, g	328	,,	...	,, g	359	W. of 2027	o
2029*	Corderilla Mts. (sic)	...	* ,, ? (XX.-9 aa) ?	...	Cordillera Mts.	358	...	Λ
2030*	Corderilla A	...	*(Near) Rocca (?) a	...	Rocca A	...	S.E. of Roc.	Λ
2031*	,, B	...	*(Near) Crüger (?) a	...	Crüger B	...	(Far) . of 2030	Λ
2032	,, γ	...	(Not named)	...	Rocca γ	...	S.E. of 2030	Λ
2033*	,, δ	...	*Crüger or Rocca a (2)	...	,, δ	...	W. of 2032	Λ
2034	Eichstädt	396	Eichstädt (XX.-9)	...	Eichstädt	361	...	o
2035	,, a	...	(Not named)	...	,, a	361	N. of Eich.	o
2036	,, b	...'	,,	...	,, b	361	N. of 2035	o
2037	(Not named) (see 2073)	...	,,	...	,, c	...	W. of 2035	o
2038	Eichstädt d	...	,,	...	,, d	...	S.W. of Eich.	o
2039*	,, a	397	* ? Eichstädt a (not named in map)	30 & 35	,, a (1)	361	S. of Eich.	Λ
2040	,, β	...	Eichstädt β (1)	...	,, β	...	N.E. of Eich.	Λ
2041	(Not named)	...	,, ❧ β (2)	...	(Not named)	...	S.E. of Eich.	Λ
2042	Eichstädt B	396	,, B (not named in map)	35	Eichstädt B	360	N.W. of Eich. (in 2081)	Λ
2043	,, δ	396	(Not named)	...	Crüger (?) δ	...	S.E. of 2042	Λ
2044	,, ε	...	,,	...	Eichstädt ε	360	W. of 2038	Λ
2045	Rook Mts.	397	Rook Mts. (XX.-9 ββ) (2040 & 2041)	...	Rook Mts.	358	...	Λ
2046	Byrgius	395	Byrgius (XX.-8)	269	Byrgius	360	...	o
2047	,, A	395	Byrgius A	30	,, A (1)	360 (4th par.)	W. of Byr.	o
2048	,, B	395 & 399	,, B	...	,, B	360 & 362	W. of 2047	o

Fig. 98. A typical page from Blagg's *Collated List of Lunar Formation's*, illustrating some of the difficulties she encountered in correlating features on the three maps.

MAP OF THE MOON   SECTION III.

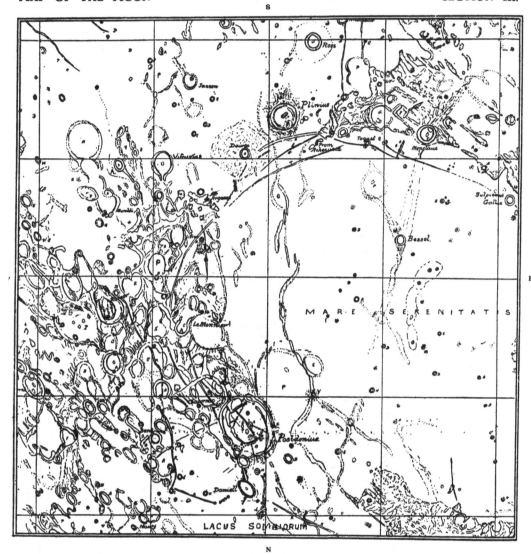

Drawn by WALTER GOODACRE, F.R.A.S., 1910.

tuted to deal with various aspects of astronomy such as Time, Asteroids, Stellar spectra, Ephemerides and so on. Commission 17 was formed to deal with lunar nomenclature, with H. Turner as President, and M. Blagg, G. Bigourdan, W. Pickering, and P. Puiseux as members.

The report of the first meeting of Commission 17, which was held in 1922, gives what turned out to be a rather over-optimistic view of the

Fig. 99. Section III of Goodacre's map (1910), showing his use of simple form lines to depict features.

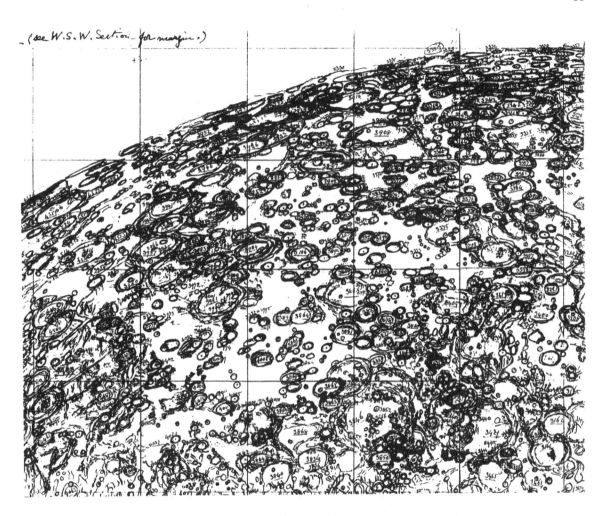

(see W.S.W. Section for margin.)

Fig. 100. Section 1 of Blagg's early MS map, with *Collated List* numbers included.

status of the project. Miss Blagg reported that she had added feature numbers from her *Collated List* (CL) to the four central sections of the map as completed by Wesley, and had drawn the ten outer sections and added the CL numbers to them as well (fig. 100).

The report of the next meeting (1925) reveals that the Committee had been a bit too hasty in assuming that most of the donkey work had been completed; Blagg produced a list of suggestions for future procedure that needed to be discussed by the Committee before further progress could be made. At the next meeting (1928) she produced some sizable lists of names for which choices needed to be made. The situation had become notably more complicated since publication of the CL because other authors of maps or drawings had added many new names. Thus she lists additions by

Birt, Gaudibert, Lee, Pickering, Elger, Müller, Lamèch, Wilkins, and more than 70 by Krieger! Of course, some of these authors had not done their homework, and further cases of two names on one feature, or duplications, or sloppy labelling came to light. These are dealt with briefly at the end of this chapter.

FINALLY, AN INTERNATIONALLY ACCEPTABLE DOCUMENT

Together with the able assistance of Karl Müller (1866–1942), a retired Czechoslovakian government official and amateur astronomer with a longtime interest in lunar matters, it took another seven years to sort out the confusion and to see the final outcome of 30 years' work in actual printed form. This was titled *Named Lunar Formations*, by Mary A. Blagg and K. Müller, and consists of two parts: Volume 1, Catalogue; and Volume 2, Maps. Vol. 1 lists over 6100 formations, each with its CL number or supplement, agreed-upon designation, rectangular (i.e. xi/eta) coordinates, diameter in terms of thousandths of the lunar radius for circular features, the serial number in Franz's or Saunder's position catalogue, and the authority for the designation, with earlier names or designations given for completeness.

Mädler's scheme of using capital and lower case letters to distinguish between measured and unmeasured positions was not continued, capital Roman letters being used exclusively for craters and valleys, and lower case Greek letters for elevations. Rilles are designated by Roman numbers, e.g. Triesnecker IIIr. A number of quite small craters, whose positions had been determined by Saunder or Franz because of their sharpness or brightness, were given new letter designations. This led to the peculiar situation in which most large and many small craters have designations, while numerous craters of intermediate size remain anonymous.

The map itself is in 14 sections, the central four squares having been very carefully drawn by Wesley, as already noted, at an original scale of 50 inches to the lunar diameter, but reduced to 36.5 inches in the atlas. Each section includes the mean centre of the Moon's visible face, i.e. the point where the Moon's Equator intersects its Prime Meridian, and is six-tenths of the lunar radius in both width and height, with a small overlap on all sides (fig. 101).

Blagg drew the remaining ten sections in pencil on metric graph paper at a scale of 1 m to the lunar diameter, tracing them from her original drawings, but the final printed scale varies from 30 to 33 inches to the diameter. She wrote, mostly with a somewhat uncertain hand, all the

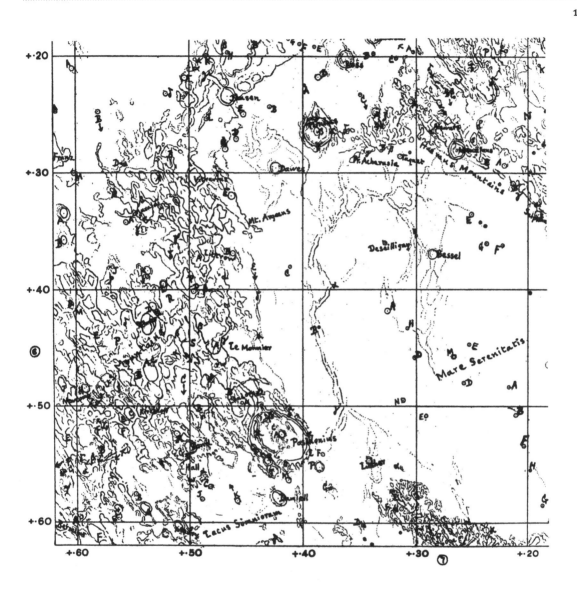

Fig. 101. Part of section 5 of the IAU map (1935) based on Wesley's drawing with Blagg's handwritten names and letters added.

names and letters in ink (over 6100 as already noted). Unfortunately the use of pencil on graph paper marked with millimetre squares does not result in the clearest of maps, especially for the Moon's limb regions where foreshortening crowds the surface features into restricted space (fig. 102). However, to be fair, we must quote her own assessment of her drawings, which, she writes:

> . . . do not profess to be more than a stop-gap, – sufficiently clear and accurate to enable the Committee to find in them the various objects named or

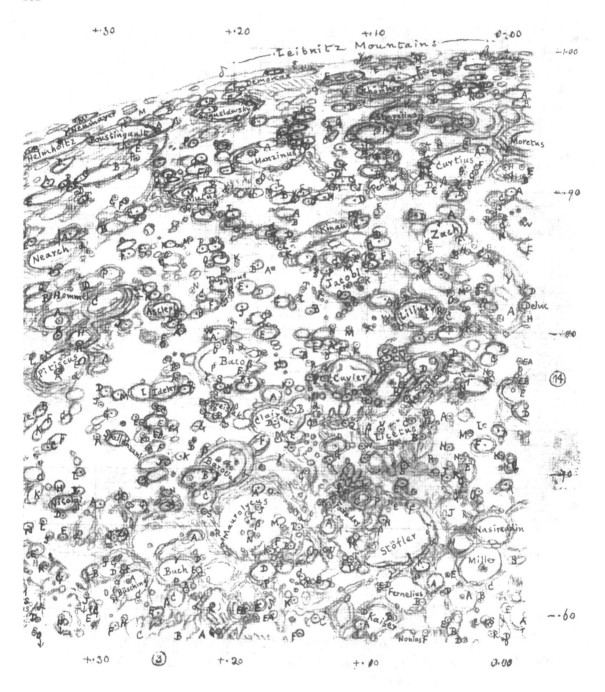

Fig. 102. (opposite) Part of section 1 of the IAU map which was drawn and lettered by Blagg. The fuzzy pencil lines, somewhat uncertain letters, plus millimeter graph paper background introduce difficulties in reading this map.

CL.No.	Designation		Co-ords.	Diam.	Sa.No.	Fr.No.	Authority.
643	Groves β		+401 +651	Λ	..	..	S(b)
644	Plana		+348 +672	21	..	..	M
645	Plan. C		+335 +679	8	2460	479	M
645a	..	D	+329 +666	4	2441	..	Com.
645b	..	E	+304 +650	4	2359	..	Com.
645c	..	F	+312 +640	3	2377	..	Com.
646	..	α	SSWw	Λ	..	..	N
646a	..	δ	+290 +631	Λ	2324	..	Com.
646b	..	γ	+288 +634	Λ	2318	..	Com.
646c	..	β	+291 +638	Λ	2325	..	Com.
646d	..	ε	+297 +627	Λ	2343	..	Com.
647	Bürg		+334 +708	18	2458	..	M
648	Bürg A		+373 +729	8	2563	354	M
649	..	B	+304 +678	4	..	..	M
650	..	α	NEw	Λ	..	..	M
651	..	β	+286 +708	Λ	..	..	M(B)
652	..	γ	NE edge 658	Λ	..	..	N
654	..	Ir	nNE to S 659	/	..	..	N(ζ)
658	Lacus Mortis		+280 +680 to +410 +770	+	..	..	Ric.
659	Baily		+328 +763	0 17	..	..	M
660	Baily A		+344 +751	11	2490	344	M
661	..	B	+362 +776	4	2535	..	M
662	..	C	+350 +768	5	..	..	M
*663	..	D	+327 +766	0 8	..	..	M
664	..	α	Ew	Λ	..	..	M
665	..	β	+308 +770	ΛΛ	..	..	M
666	..	ε	N Bürg	ΛΛ	..	..	N(Γ)
667	..	γ	S 665	Λ	..	..	M
668	..	δ	+300 +730	Λ	..	..	M
*669	Gärtner		+290 +855	0 60	..	..	Schr.
670	Gärt. A		+292 +872	5	..	..	M
*671	..	B	+320 +850	4	..	..	M(b)

Fig. 103. A typical page from the IAU Catalogue that accompanies the map.

663   A triangular bay on inner NEw is substituted for M's minute crater.
669   An incomplete formation between 669 and 696 is called Danjon by Lamèch.
671   M's b: N's is very minute and further north.

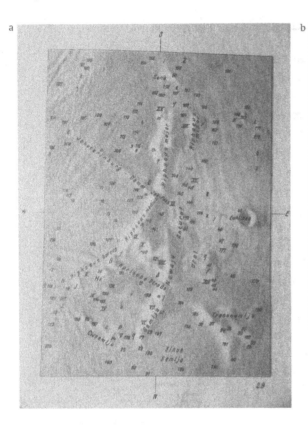

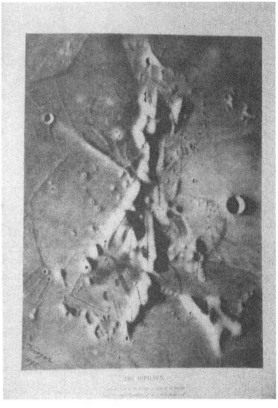

lettered by the three authorities: – but not aiming at any sort of equality with the central sections, either in artistic merit or exactitude. 1922 Mar 24.

As so often happens with stop-gap measures or items, they become permanent fixtures, and so it was that 1935 saw the publication of *Named Lunar Formation's*, or NLF as it is usually known, with Mary Blagg's unimproved maps and lettering. She possibly was chagrined at this, but could bask in the satisfaction that her indefatigable efforts had resulted in an internationally approved and accepted document that had averted chaos in the rather esoteric subject of lunar nomenclature (fig. 103).

At all events, the NLF catalogue and maps constituted an acceptable compromise between the various nomenclatures that had been published during the previous hundred years and, being an official document of the IAU, was adopted internationally. Its immediate task completed, IAU Commission 17 was in 1938 renamed 'Movement and Figure of the Moon', and its members were absorbed into Commission 16 (Commission for

Fig. 104. Krieger's map of Montes Riphaeus, based on a Lick photograph. He has added several new names and a large quantity of numbers and letters, which are printed on a translucent paper overlay (a). The enlargement from the original negative is shown in (b), with his added fine details drawn directly onto the photo.

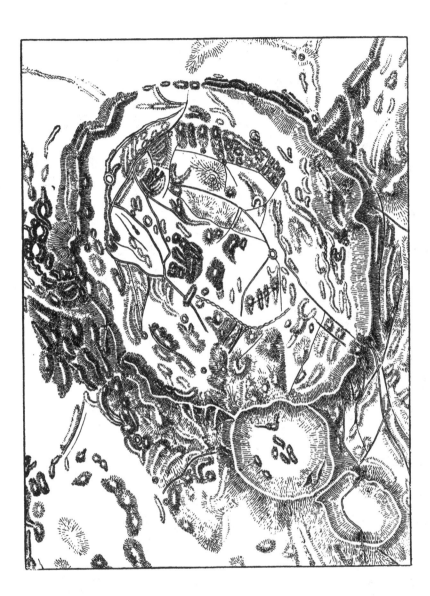

Fig. 105. Fauth's map of the crater Gassendi, which reverts to the hachure method for depicting slopes.

Physical Observations of Planets and Satellites), which would deal with any future issues concerning lunar nomenclature.

Before passing on to events after the Second World War, and the upheavals in the subject engendered by the advent of the Space Age, we need to return to the earlier years of the twentieth century and take a brief look at the chief 'perturbers of the nomenclatural status-quo', together with one or two other noteworthy milestones in lunar mapping. The major 'perturber' was Johann N. Krieger (1865–1902), a Bavarian astronomer, who

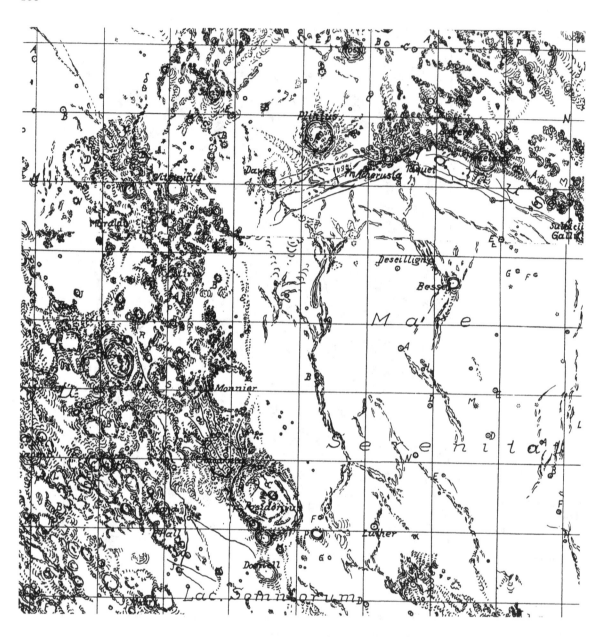

Fig. 106. A portion of Fauth's smaller General Chart showing a somewhat different method of portrayal, and covering the same area as figs. 76, 82, and 99.

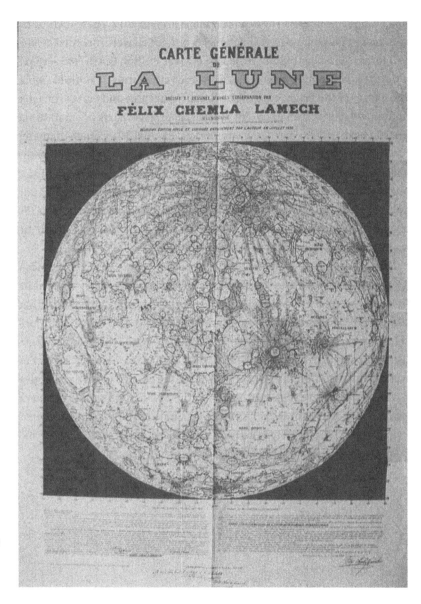

Fig. 107. Lamèch's lunar map of 1956. Earlier versions included numerous new names, the majority of which were not accepted by the IAU.

emulated Schröter to the extent that he observed many limited areas of the lunar surface, but rather than attempting to make drawings, started with enlargements of Paris Atlas photographs and entered the finer details on those. As with Schröter, he was obliged to add names to the images in order to describe the various features that he observed (fig. 104). The first part of his work was published in 1898, with two more volumes posthumously published by R. König (1865–1927) in 1912. These include no less than 73 new names (see Appendix N), but not all were included in NLF.

Another significant figure from this era was Philipp Fauth (1867–1941), a German astronomer who espoused some rather strange notions about the nature of the bright tops of the lunar mountains and crater rims (ice and snow!), but was an excellent cartographer. His charts of small lunar areas are expertly executed in hachures (e.g. fig. 105). His 34-inch-diameter map in six sections, which was published in 1936 along with his book *Unser Mond* (*Our Moon*), uses a novel type of representation for the topographic features (fig. 106) that is aesthetically quite pleasing. It does not include any albedo shadings, and is plotted on the rectangular xi/eta grid, thus following the lead of Goodacre and the IAU. He added seven new names to the map, taken from his earlier charts (Appendix O).

The other main 'nomenclature perturber' was Felix C. Lamèch (1894–1962), a French astronomer who built Corfu Observatory and was its first director, and who published many charts of various lunar areas that illustrated the numerous 'aires elliptiques', obscure and arguable elliptical features first pointed out by G. Delmotte (1876–1950). He also published a colourful map of the entire disc with a diameter of 24 inches which illustrates Delmotte's other idea that the crater Tycho is a directive centre and an alternative south pole on the Moon (fig. 107). He added at least 50 new names over the pre-NLF years but only eight were accepted (Appendix P). The very few additions made by Blagg, Müller, and the Committee are listed in Appendix Q.

FOURTH ERA

THE SPACE AGE DEMANDS CHANGES

CHAPTER 10

....

SETTING UP GUIDELINES

THE IAU REJECTS SOME PROPOSED NEW NAMES
At the first General Assembly of the IAU (in 1948) following the Second
World War, H.P. Wilkins (1896–1960), a British amateur selenographer, put
forward a proposal to accept 22 new names that he wished to include on
his 100-inch map. Commission 16 members turned this down, noting that

> The formations to which he wishes give names are, in general, small or
> observable near the limb, in mediocre conditions. Most of them are
> already designated by letters, which were adopted by the IAU as definitive.

However, E. Antoniadi's 1942 suggestion to name '. . . the large rectangular
depression situated south of the Straight Wall, which contains the craters
Hell and Lexell . . .' for **Henri Deslandres** was passed.

Further proposals by Wilkins were turned down at the 1952 and 1955
IAU General Assemblies for the same reason, the following resolution
being adopted unanimously by Commission 16.

> The Commission recommends to the Union that at the present time – and
> particularly pending the completion of the proposed photographic map of
> the Moon – no official recognition shall be given to additional lunar
> nomenclature.

Despite this he included almost 100 unofficial new names (see Appendix
R) and numerous extra letters in his (then) recently published 100-inch
Moon map. This map follows the 25 section plan adopted by Lohrmann,
Schmidt and Goodacre. Unfortunately, his cartographic abilities did not
match his infectious enthusiasm for the subject, and the smaller details in
his crowded map are not trustworthy (see fig. 108). Two names from his 60-
inch map of 1926 were included in NLF (**Goodacre** and **Mee**).

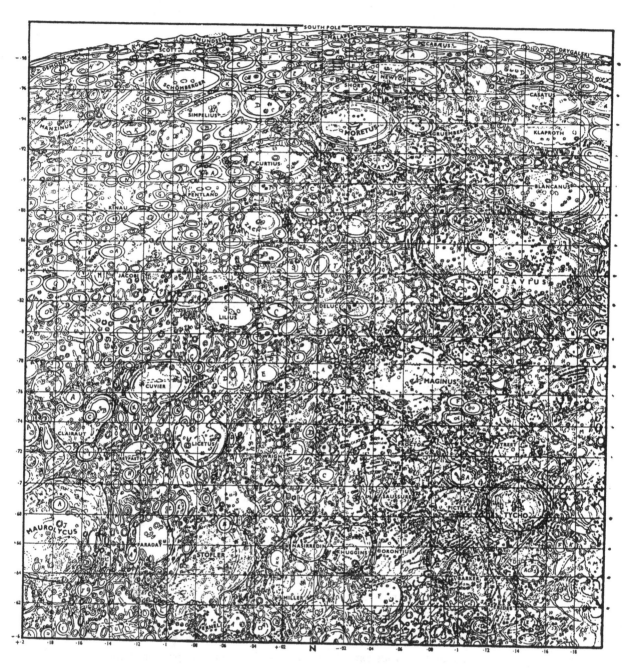

Fig. 108. Section 23 of Wilkins's 200-inch lunar map. The addition of fictitious fine detail has so cluttered the map that it is virtually uninterpretable.

THE WINDS OF CHANGE ARE A'BLOWIN'

The 'photographic Moon map' mentioned above was proposed by the President of Commission 16, Gerard P. Kuiper (1905–73), Director of Yerkes and McDonald Observatories and one of the very few professional astronomers interested in lunar research at that time. Such was the overwhelming disinterest in the Moon that I was the only one to reply to his proposal!

Some inconsistencies, errors and other problems in NLF, often traceable back to the 1913 CL, began to show up in studies made during this same period, and especially during preparatory work for the *Photographic Lunar Atlas* (PLA) and *The System of Lunar Craters* (SLC), two projects that Kuiper initiated at Yerkes Observatory in 1958 as a follow-up to his 1955 proposal. I worked on the first of these two projects, while D.W.G. Arthur, a British Ordnance Survey cartographer and amateur selenographer, was in charge of the second. Table III in the booklet accompanying the PLA lists some 45 minor changes to spelling and typography, 13 deletions of minor or unidentifiable features, and about 18 clarifications of outline or identification. These changes, as well as some considerably more far-reaching issues, were considered prior to and during the 1961 IAU General Assembly.

THE SPACE AGE LOOMS

The 1960 success of the Soviet *Luna 3* mission in photographing part of the Moon's far side, resulting in a map containing 18 names of features never before seen, brought into sharp focus the lack of specific rules for naming such newly revealed markings. It became clear that eventually the back of the Moon would be photographed in detail comparable to that of the near side. Since nearside maps are normally produced with south up, to accommodate the inverted image presented to observers viewing from Earth's northern hemisphere, how should maps prepared from spacecraft imagery be oriented? And what about 'east' and 'west', which are 'sky' directions and thus opposite to true directions on the Moon?

The result of these considerations was the formulation of multi-part resolutions, which are reproduced in full in Appendix S. Briefly, these were that 'east' and 'west' would now correspond with terrestrial mapping, and that astronautical maps would be oriented north up; newly named craters etc. must be of people deceased; and feature types, e.g. mount, rille, valley etc. must be in Latin, thus: mons, rima, vallis. Incidentally, as I mentioned earlier, the word 'rille' was introduced by Schröter, meaning a groove. Either Neison or Birt mis-translated this into

English as 'rill', which is a narrow mountain stream – a not very appropriate term for the water-less Moon. Nevertheless the word persisted in English literature until the late 1950s, when G. Fielder, a British selenographer, successfully advocated returning to the German spelling.

Because of the then current considerable increase in lunar studies and the onset of exploration by spacecraft, a Committee (16a) for Lunar Nomenclature and Cartography was formed at the 1961 IAU Assembly consisting of A. Dollfus, Z. Kopal, K. Koziel, G. Kuiper, D. Martynov, A.A. Mikhailov, and M. Minnaert.

THREE BUSY YEARS, 1961–4

Two major projects affecting nomenclature were already under way at this time: 1) a complete revision of NLF (the SLC under Arthur's direction as mentioned earlier, and 2) the production of a rectified lunar atlas (i.e. 'astronaut's eye' or vertical views of all areas of the lunar near side. A comprehensive programme of producing the *Lunar Aeronautical Chart* (LAC) series of 1:1 million scale maps by the US Air Force Chart and Information Center (ACIC), St Louis, implemented the updates given in those two projects. The production of the *Rectified Lunar Atlas* (RLA) dramatically emphasized the shortage of names and letter designations in the limb regions once the foreshortening had been removed, as did also the LAC charts of those areas. Accordingly, Arthur and I added 66 names (50 new), thereby providing a workable density of names near the limb (Appendix T).

All three publications (SLC, RLA, and LAC) were exhibited and discussed at the 1964 IAU General Assembly. The 66 names, plus a number of other changes such as latinizations (as recommended by the 1961 Assembly), a few deletions and additions etc. were adopted as official. An interesting side issue was to choose a name for the semi-enclosed mare area, chosen by Whitaker before the mission as the optimum impact target point for NASA's *Ranger 7*, which the spacecraft indeed photographed in ever-increasing detail until it crashed there. Kuiper suggested a choice between **Mare Exploratum** or **Mare Cognitum**. After considerable discussion, the second of these (the Known Sea, or The Sea That Has Become Known) received the vote. Commission 17 was renamed 'The Moon', and nomenclature activities now came under its aegis.

1964–7 – THREE EVEN BUSIER YEARS

The SLC catalogue and map project was completed in 1966, and was exhibited at the 1967 IAU General Assembly, at which 'Arthur and Kuiper

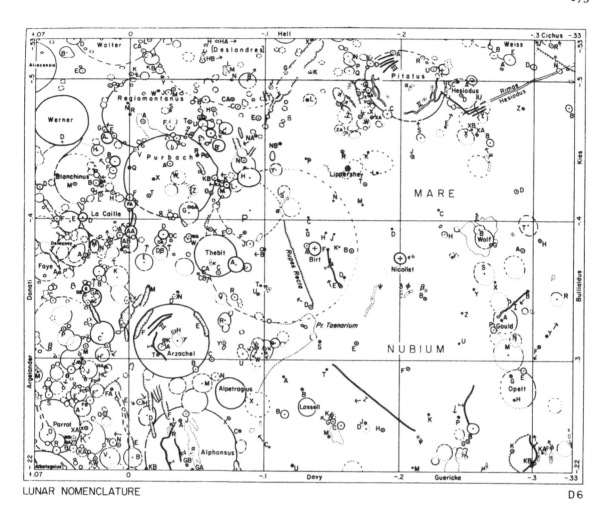

LUNAR NOMENCLATURE   D 6

Fig. 109. A section of
the outline map that
accompanies the *System
of Lunar Craters* catalogue.
It shows the officially
accepted IAU nomen-
clature as of 1967.

reported completion of the scheme of lunar nomenclature for the Moon's
visible hemisphere, which was proposed and approved by the IAU at its
1964 Hamburg meeting.' These documents represented a complete revision and extension of the NLF (1935). The catalogue lists only craters,
giving a reference number, NLF number, name or designation, xi/eta/zeta
and lat./long. coordinates, diameter in thousandths of the lunar radius
and kilometres, sharpness of the rim, and data on central peaks and type
of background.

The included 44 sectional maps (fig. 109) give also a revised version of
the remainder of the nomenclature (maria, montes, rimae etc.). They are
outline maps only, and include very little beyond named or designated

features. They were later combined into four quadrant maps (à la Mädler) with a scale of 1 m to the lunar diameter. Mare shadings were also added. The quadrants were published under the title *Lunar Designations and Positions* (LDP) and are still (as of 1998) available commercially. However, the limited production of the catalogue prevented its becoming universally available, even though it was officially approved by the IAU.

Recapitulating, the situation in 1967 was that SLC and LDP were adopted as representing the officially approved lunar nomenclature, supplanting NLF with its various errors. Those maps of the LAC series that had been published by then adhered to the SLC nomenclature, and were thus 'official' in that sense. New names given to several craters in RLA that lay on or just beyond the mean limb and thus excluded from SLC were also included in the approval, as were the *Luna 3* names from 1961. Fig. 110 illustrates a typical page from SLC.

ZONDS AND *ORBITERS* CALL THE SHOTS

Those of us in the lunar photography/cartography/nomenclature field were pleased to see the results of our labours receiving general approval, so we were not quite prepared to deal there and then with the new curve thrown at us by Soviet astronomers, who had brought along to the 1967 Assembly a number of newly printed volumes of a book, *Atlas Obratnoi Storony Luny, Chast 2* [*Atlas of the Far Side of the Moon, Part 2*]. This was based on the earlier *Luna 3* images plus those taken by the *Zond 3* spacecraft, which had photographed most of that portion of the Moon's far side not covered by *Luna 3*. The accompanying map and list of new names caused not a little consternation at the time, not only from the *fait accompli* nature of the operation, but also from the makeup of the list, about 45% of which was of Russian names!

Also exhibited were photographs from *Lunar Orbiters*, especially the high-resolution photographs of the Moon's near side from *Orbiter 4*, which were assembled as a mosaic on a hall floor and covered with transparent plastic so that people could walk on them (shoeless!). ACIC also exhibited preliminary maps of the lunar far side, compiled from the incomplete coverage of *Orbiters 1–4* (*Orbiter 5* was filling in the gaps at the time of the Assembly!).

The Committee for Lunar Nomenclature and Cartography added F. Heyden, D. Menzel, and E. Shoemaker as members, Kopal being chairman. After considerable discussion on how best to name the newly revealed (and still to be imaged) farside features, it was generally agreed that the

Fig. 110. (opposite) A typical page from the SLC catalogue, giving positions, diameters, rim sharpness, central peak and background data.

16

Ref.	B & M	Designation	ξ	η	ζ	λ	β	D	K	C	B	c.z.
13499			+ .391	+ .491	+ .778	+ 26.6	+ 29.4	65.84	114.44	5f	aM	0
13518	(490C)	Luther H	.313	.588	.746	22.7	36.0	4.06	7.06	1	C	0
13539		Luther X	.333	.590	.736	24.3	36.1	2.25	3.91	2	pM	0
13544	491	Luther	.342	.547	.764	24.1	33.1	5.48	9.53	1	pM	0
13577	490B	Posidonius G	.376	.570	.731	27.2	34.7	2.62	4.55	1	pM	0
13584	490A	Posidonius F	.383	.541	.749	27.0	32.7	3.45	6.00	1	pM	0
13585	490	Posidonius P	.385	.553	.739	27.5	33.5	8.69	15.10	1	pM	0
13603		Plana G	.302	.630	.715	22.8	39.0	5.35	9.30	1	pM	0
13604	645B	Plana E	.304	.650	.696	23.5	40.5	3.65	6.34	1	pM	0
13610		Luther K	.314	.608	.729	23.2	37.4	2.01	3.49	1	pM	0
13610A			.312	.609	.729	23.1	37.5	2.01	3.49	1	pM	0
13614	645C	Plana F	.312	.640	.702	23.9	39.7	2.69	4.68	1	pM	0
13621		Luther Y	.325	.617	.717	24.3	38.0	2.06	3.58	1	pM	0
13626	645A	Plana D	.329	.666	.669	26.1	41.7	4.19	7.28	1	pM	0
13637	645	Plana C	.335	.679	.653	27.1	42.7	7.91	13.75	1	pM	0
13640	467	Daniell D	.348	.602	.719	25.8	37.0	3.64	6.33	1	pM	0
13640A			.345	.604	.718	25.6	37.1	2.02	3.51	2	pM	0
13657	644	Plana	.350	.672	.653	28.2	42.2	25.46	44.25	3	aMC	P
13666	639A	Mason B	.369	.667	.647	29.6	41.8	6.27	10.90	1	C	0
13668	639	Mason A	.368	.680	.634	30.1	42.8	2.83	4.92	1	pMC	0
13676			.374	.666	.645	30.0	41.7	2.76	4.80	1	C	0
13677	638	Mason	.373	.678	.633	30.4	42.6	18.72 24.80	32.54 43.11	3f	C	0
13693			.391	.633	.668	30.3	39.2	2.46	4.28	1	pM	0
13710		Lacus Mortis	.319	.708	.630	26.8	45.0	93.42	162.38	4f	aMC	0
13716			.314	.763	.565	29.0	49.7	5.44	9.46	2	C	0
13718		Baily K	.316	.782	.537	30.4	51.4	1.95	3.39	1	pM	0
13723			.320	.733	.600	28.0	47.1	2.01	3.49	1	pM	0
13726	659	Baily	.327	.763	.558	30.3	49.7	15.40	26.77	4f	aMC	0
13730	647	Bürg	.334	.708	.622	28.2	45.0	22.76	39.56	1	pM	P
13745	660	Baily A	.344	.751	.564	31.3	48.6	9.95	17.29	1	pM	0
13766			.360	.769	.528	34.2	50.2	2.75	4.78	2	pM	0
13766A			.363	.762	.536	34.0	49.6	2.86	4.97	1	pM	0
13766B			.368	.763	.531	34.7	49.7	3.18	5.53	2	pMC	0
13767	661	Baily B	.362	.776	.517	35.0	50.8	4.21	7.32	1	pM	0
13772	648	Bürg A	.373	.729	.574	33.0	46.8	7.13	12.39	1	pM	0
13773			.378	.730	.569	33.5	46.8	2.91	5.06	2	pM	0
13775			.370	.758	.537	34.5	49.2	2.90	5.04	2	pMC	0
13776			.379	.765	.521	36.0	49.9	19.83	34.47	4f	aMC	0
13808			.307	.881	.360	40.4	61.7	6.88	11.96	3	C	0
13809		Schwabe E	.301	.899	.318	43.4	64.0	10.95	19.03	3	C	0
13817			.319	.877	.359	41.5	61.2	6.44	11.19	2	C	0
13818			.313	.881	.355	41.4	61.7	6.23	10.83	2	C	0
13818A			.313	.882	.352	41.6	61.8	3.44	5.98	2	C	0
13826		Gärtner G	.325	.863	.387	40.0	59.6	12.77 18.91	22.20 32.87	4f	aMC	0

traditional system as used on the near side should be extended rather than resorting to alphanumeric or other similar schemes. The merits of expanding the categories to include cosmonauts, astronauts and others was also touched on. As a result of all this, the following resolution was adopted:

> The assignment of names and permanent designations to features on the far side of the Moon will be postponed until the 14th (i.e. 1970) General Assembly. As an interim measure, the WGLN will assign numbers to about 500 major lunar formations.

EXTENSION TO THE FARSIDE, 1967–71

A sub-group of the Nomenclature and Cartography Committee was formed consisting of D. Menzel (chair), Mikhailov, Minnaert, and Dollfus, '... none of whom is personally and directly engaged in the work of Lunar Topography.' This later turned out to be a disastrous philosophy.

This four-year period saw a great deal of activity and correspondence devoted to the production of an acceptable farside nomenclature. Several different schemes were proposed and discussed, as were numerous individual names, submitted by representatives of many member nations of the IAU, and generic terms for the newly identified 'basins'. A constant barrage of correspondence ensued between the Chairman of the WGLN, Commission 17 members, and numerous other individuals and organizations who had, or in some cases thought they had, some connection with or claim to the whole process of farside names.

One particular point of contention was the application of the names of cosmonauts, both deceased and alive, on features, as had been done in the new Soviet atlas. Names of living people were not allowed because of the 1961 resolution, yet here were pioneers worthy of some recognition. Agreement was finally reached by choosing nine Soviet cosmonauts and nine US Astronauts, six living and three deceased in each case. Eight of the cosmonaut names were given to craters in or near, very aptly, Mare Moscoviense, while Gagarin, the first man to orbit Earth, names a large crater at the same longitude. Of the Astronauts, three living and three deceased (in the *Apollo* capsule fire) have craters, also very aptly, in or near the Apollo Basin, while those on the first Moon landing mission (*Apollo 11*) have their names on three small craters near the landing site, which received the name 'Statio Tranquillitatis'.

This correspondence proved to be tedious but effective, allowing many compromises to be reached, and also preventing the Chairman from

implementing two or three plans that contravened well-established cartographic and nomenclatural principles, e.g. names to be arranged alphabetically by latitude zones.

A preliminary list of names, and some changes etc., was available, along
with a hastily prepared farside map by ACIC with the new names included,
at the 1970 IAU General Assembly. Some unacceptable changes soon came
to light – names that had been applied to some limb craters and approved
in 1964 were moved to other craters! Here was the first case of the trouble
that can stem from people who are not 'personally and directly engaged in
the work of Lunar Topography' getting into a field in which they have no
experience. Some vigorous protests backed up by relevant photographs
caused the WGLN chairman to retract the changes.

In 1971 a definitive list of farside nomenclature, together with a few
additions and changes to the nearside, was published under the title
'Report on Lunar Nomenclature' in *Space Science Reviews*. This list was generally acceptable to all users, the only remaining problem being that the
base map which was originally used for the allocation of crater names
(ACIC charts LMP 1, 2, 3; 1st edn.) contained several poorly represented
craters, so that a few names were given to scarcely identifiable features,
while a few prominent craters remained anonymous. The Introduction
documents the chief decisions that went into the production of the
Report. Subsequent review revealed a number of errors in the Report, such
as incorrect coordinates, missing names etc.

CHAOS SETS IN, 1972–3

Following the eminently successful metric (mapping) camera photography from *Apollo 15, 16*, and *17*, NASA decided to prepare a series of 1:¼
million scale lunar orthophotomaps (LO) and lunar topographic orthophotomaps (LTO), the former being rectified and enlarged metric camera
images, and the latter being the same but with added contours and
nomenclature. As a result of an unwritten decision back in 1966 to accept
IAU nomenclature and the rules pertaining to it, NASA now established a
direct link with the IAU for cooperation on nomenclature matters,
dealing directly with WGLN through its chairman (Menzel). Thus the
WGLN non-experts now found themselves as prime advisors to NASA for
questions of nomenclature and cartography, the NASA official in charge of
cartography programs delegating many of the decisions to WGLN!

The result of all this was that impractical rules were set up for naming
map sheets; these rules were confused with rules for nomenclature;

revolutionary new rules for nomenclature were proposed and implemented, while traditional conventions and ethics were disregarded. Written protests from users proved to be fruitless, being ignored by NASA (cartography section) and rejected by the WGLN. The financial inaccessibility of the 1973 (Sydney) IAU General Assembly to most lunar mappers and map users, and indeed to most Commission 17 members, as well as the known suppression of letters of protest from selenographers, ensured the adoption of the new proposals.

As an example of the crater name/map sheet name problem, sheet LTO 61C2 is named after the small crater 'Cameron' (traditionally named Taruntius A), even though it contains the entire crater Taruntius. The reason that it could not be given that name is that Taruntius is the name of the drawn (airbrushed) 1:1 million scale map of the area!

Another resolution sought eventually to replace all letter designations (over 5500) with names, a mind-boggling prospect that scarcely merits a second thought. Yet another signalled the future use of the names of 'writers, painters, musicians etc.', thereby contravening the first part of Resolution 2 from 1961.

Some unexpected changes occurred in the spelling of a few names from the form in earlier lists, notably of Russian names transliterated from the original Cyrillic orthography. Thus the name **'Hansky'**, the familiar form, now became **'Ganskij'**, the spelling favoured by the IAU transliteration system, which reads as 'Ganskidge' to all English speaking users. Similarly, the name **'Van Gu'**, which baffled most, if not all Westerners (especially Dutchmen!), is actually **'Wan-Hoo'**, rather well known as an early Chinese rocket experimenter.

....

CHAPTER 11

PLANETS AND SATELLITES
SET THE RULES

FURTHER DEVELOPMENTS, 1973–6

With the acquisition of ever growing amounts of imagery of the planets at that time, it became clear that committees would be required to deal with the nomenclature problems of the newly revealed surfaces, just as for the Moon six years earlier. The Working Group for Planetary System Nomenclature (WGPSN) was therefore formed, to coordinate the work of several Task Groups which covered the planets and satellites, with a particular view to preventing a recurrence of the Mars situation where 134 out of 189 names chosen for Martian craters were already in use on the Moon.

The Task Group for Lunar Nomenclature (TGLN), as it was now named, continued its wayward policies, making things even worse by removing NLF, SLC, and LAC chart letter designations, naming rilles and ridges for composers, poets and writers, moving long-established names to different features, and so on. As an example of this last travesty, the names **Lyell**, **Franz**, and **Hypatia** were moved to the nearby craters **Proclus A**, **Proclus D**, and **Hypatia** A. Such changes are the anathema of a stable, standardized nomenclature.

To complicate matters further, outside groups such as the Board of Geographic Names, geophysicists, astro-geologists, planetary cartographers, and both national and international committees all thought they should have a stake in the process. LTO charts were being published and distributed with unapproved new names, often with different spellings.

WGPSN ASSERTS ITS AUTHORITY

The Working Group convened annually after its formation in 1983 in order to keep pace with imagery being obtained from the planets, and the maps being constructed from that imagery. This helped to expedite matters when nomenclature was needed for operational and descriptive purposes, when the three-year interval between General Assemblies was impractically long.

At its very first (1974) meeting, a resolution was passed that '... names of non-scientists shall not be chosen for lunar maps.' This immediately disqualified about 60 new names used on the LTO charts that had been assigned provisionally to various craters, rilles and ridges, and even to some map sheets (fig. 111). The Chairman of TGLN protested this reversal of policy, but was over-ruled because such names were required for the new Mercurian nomenclature.

Officials in NASA, other than the very few involved in the LTO production area, were more responsive to the repeated complaints of those of us who were directly involved in lunar cartographical matters. The ongoing publication of the LTO charts with their ever-changing nomenclature was causing chaos in those areas covered by the charts. Cataloguers, cartographers, researchers, and especially curators of lunar imagery and cartography collections whose computerized retrieval systems had to be constantly updated and expanded, were unanimous in their condemnation of the new nomenclature.

Accordingly, in 1974 NASA formed a users' group (Lunar Photography and Cartography Committee, LPACC), consisting of scientists and cartographers long involved in lunar matters, to provide sensible guidelines for the mapping programmes. This later expanded to include planetary photography and cartography (LPPACC) in order to prevent a recurrence of the lunar fiasco. WGPSN lent a sympathetic ear to the recommendations of LPPACC and kept a tighter rein on the unilateral actions of the Task Group (TGLN). Thus at their 1975 meeting, WGPSN recommended that maps containing approved new names must include, in brackets and where appropriate, the previous letter designation of the crater. The next annual meeting was held during the 1976 General Assembly of the IAU, when a large list of new names was approved, and was published in the Transactions of the IAU for that year.

STABILITY BEGINS TO RETURN

The chairman of TGLN (D. Menzel) died shortly after the 1976 IAU Congress and was replaced by P. Millman, who was also chairman of WGPSN at that time. Cooperation between WGPSN, TGLN, and LPPACC at last ensured progress in resolving various outstanding issues. Thus it was agreed that

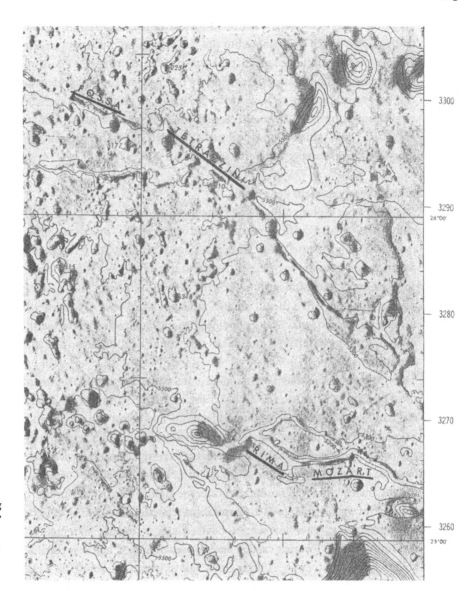

Fig. 111. A portion of
LTO chart 41A3 showing
rejected names 'Mozart'
and 'Tetrazzini' (under-
lined here). The generic
term 'fossa' was also
rejected in 1976.

the designations 'GDL', 'GIRD', and 'RNII', which were applied to three
prominent farside crater chains on Soviet maps but which were not
approved by the IAU because they represented an unpopular new names
category (institutions), would now be recommended for approval, provided that they were placed in brackets under the regular 'catena' designation from the nearest named crater. So 'GDL' became 'Catena Leuschner
(GDL)'.

Another subject resolved was the transliteration of names from Cyrillic orthography, which had adhered to the IAU system up to this point. The WGPSN overwhelmingly agreed that the system used by the Board of Geographical Names was preferable to the IAU system. The chief effect of this was to replace the letter 'j' with 'y', which is much preferable for English-speaking users, and is still acceptable to other users of the Roman alphabet. Thus 'Ganskij' now became 'Ganskiy', not quite back to the long-used form 'Hansky' but quite acceptable.

LETTERS FOR FARSIDE CRATERS?

The scheme to replace all lettered craters on the Moon's near side with names was clearly quite impractical and unnecessary, and was implemented in only a limited number of cases on LTO charts. The letter designations as given in the 1935 NLF map and catalogue, but as updated in the 1966 SLC map and catalogue, were still 'official'. However, the WGPSN position was that when the IAU finally publishes a gazetteer of nomenclature for all imaged Solar System bodies, it should not concern itself with including all the lettered formations as this would overburden an already daunting task.

At this point the problem of unnamed craters on the farside of the Moon presented itself. After all, the highly cratered farside had received about 600 names, whereas the less-cratered nearside had about 800 names and 5400 letters. Many of these unnamed craters were unique or interesting in their own right, and would be the subject of investigations, research articles etc., and would need some sort of identification. The project of lettering the major unnamed landmark craters somehow fell to your present author. Not wishing to repeat the haphazard lettering scheme that afflicts the lettering on the nearside, I devised the following scheme for maximum ease in locating craters on maps: each named crater is considered to be the centre of a 24-hour clock dial in which the numbers have been replaced with Roman capital letters (I and O omitted, 24 h=Z), with Z at the north point. Thus each letter represents a fixed azimuth from the centre of the named crater, and the chosen subsidiary craters are lettered according to their closest azimuths. This scheme has been adopted for the 1:5 million scale maps of the lunar farside produced for NASA by the US Geological Survey. Figure 112 illustrates the 'clock-face' precept for letters, and a portion of the farside map with letters in place.

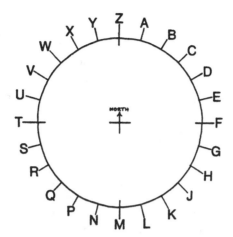

Fig. 112. 'Clock-face' scheme for lettering prominent unnamed farside craters (a). Implementation of the scheme on the USGS map of the farside is shown in (b), where arrows have been added for emphasis.

WINDING DOWN AFTER THE TURMOIL

With the termination of the production of LTO charts in 1978, and the farside nomenclature virtually settled, TGLN activities were greatly reduced from about 1979 onwards, the main items being the tidying up of loose ends left over from the previous decade, and the addition of a few names of recently deceased persons.

A major task, however, was the implementation of part 2 of Resolution 1 passed in 1976 by WGPSN

> That there be produced, for wide distribution, a comprehensive publication including all planetary nomenclature resolutions and lists of names officially approved by the IAU up to and including the 16th General Assembly; and that all future approved planetary nomenclature be published in supplementary volumes following each General Assembly.

This daunting chore was taken on by the Astrogeology Branch of the US Geological Survey, which actually was in a good position to do this work since the countless maps of the surfaces of the planets and their satellites were being prepared there. The terms of the resolution proved to be rather unrealistic, so that in the event, a single volume, *Gazetteer of Planetary Nomenclature 1994*, was finally published in 1995 that contains all names officially recognized by the IAU as of the 1994 General Congress. A page of this comprehensive catalogue is illustrated in fig. 113. Users should note that the lunar tables contain a number of errors, such as the crater Ramsden being supposedly located at the exact centre of the Moon's disc, data for Mutus crater being given for Mons Argaeus, many poorly measured crater diameters, etc. These errors had all been corrected in galley proof versions, but somehow reappeared in the final printed version.

NASA had some concerns about the omission of the lettered lunar craters from this catalogue, and in about 1979, encouraged two of us at the Lunar and Planetary Laboratory, University of Arizona, to produce a catalogue listing all the approved named and letter-designated formations as a by-product of a programme already under way. This was done, the final product being published by NASA in 1982 under the title *NASA Catalogue of Lunar Nomenclature*. It is complete up to 1981, and lists feature name, coordinates, size, and LTO map sheet number in the case of very small features. Thirtyseven new names have been approved and added since its publication; these are listed in Appendix U, while fig. 114 illustrates a sample page from the Catalogue.

Fig. 113. (opposite) A typical page from the USGS Gazetteer of Planetary Nomenclature, which does not include lettered features.

SECTION 1—NAMES LISTED BY PLANET, SATELLITE, AND FEATURE TYPE   55

MOON

Name	lat	long	diam	ct	et	quad	map	as	ad	ref	ft	origin
Helicon	40.4N	23.1W	24	EU	GR	LAC	24	5	1935	66	AA	Greek astronomer, mathematician (unkn-c. 400 B.C.).
Hell	32.4S	7.8W	33	EU	HU	LAC	112	5	1935	66	AA	Maximilian; Hungarian astronomer (1720-1792).
Helmert	7.6S	87.6E	26	EU	GE	LTO	081B3	5	1973	0	AA	Friedrich Robert; German astronomer, geodesist (1843-1917).
Helmholtz	68.1S	64.1E	94	EU	GE	LMP	3	5	1935	66	AA	Hermann Von; German doctor (1821-1894).
Henderson	4.8N	152.1E	47	EU	SC	LOC	4	5	1970	0	AA	Thomas; Scottish astronomer (1798-1844).
Hendrix	46.6S	159.2W	18	NA	AM	LMP	3	5	1970	0	AA	Don O.; American optician (1905-1961).
Henry	24.0S	56.8W	41	NA	AM	LAC	92	5	1970	0	AA	Joseph; American physicist (1792-1878).
Henry Fréres	23.5S	58.9W	42	EU	FR	LAC	92	5	1961	66	AA	Prosper; French astronomer (1849-1903); Paul; French astronomer (1848-1905).
Henyey	13.5N	151.6W	63	NA	AM	LOC	4	5	1970	0	AA	Louis G.; American astronomer (1910-1970).
Heraclitus	49.2S	6.2E	90	EU	GR	LAC	126	5	1961	67	AA	Greek philosopher (c. 540-480 B.C.).
Hercules	46.7N	39.1E	69	EU	GR	LAC	27	5	1935	66	AA	Greek mythological hero.
Herigonius	13.3S	33.9W	15	EU	FR	LAC	75	5	1935	66	AA	Herigone, Pierre; French mathematician, astronomer (fl. 1644).
Hermann	0.9S	57.0W	15	EU	SZ	LAC	74	5	1935	66	AA	Jacob; Swiss mathematician (1678-1733).
Hermite	86.0N	89.9W	104	EU	FR	LMP	3	5	1964	67	AA	Charles; French mathematician (1822-1901).
Herodotus	23.2N	49.7W	34	EU	GR	LAC	39	5	1935	66	AA	Of Halikarnassus; Greek historian (c. 484-408 B.C.).
Heron (Hero)	0.7N	119.8E	24	AF	EG	LTO	065D3	5	1976	0	AA	Egyptian inventor (unkn-c. 100 B.C.).
Herschel	5.7S	2.1W	40	EU	GB	LTO	077A3	5	1935	66	AA	William; British astronomer (1738-1822).
Hertz	13.4N	104.5E	90	EU	GE	LOC	3	5	1970	0	AA	Heinrich R.; German physicist (1857-1894).
Hertzsprung	2.6N	129.2W	591	EU	DE	LOC	1	5	1970	0	AA	Hertzsprung, Ejnar; Danish astronomer (1873-1967).
Hesiodus	29.4S	16.3W	42	EU	GR	LAC	94	5	1935	66	AA	Hesiod; Greek humanitarian (c. 735 B.C.).
Hess	54.3S	174.6E	88	NA	AM	LMP	3	5	1970	0	AA	Victor F.; American physicist (1883-1964); Harry H.; American geologist (1906-1969).
Hevelius	2.2N	67.6W	115	EU	PO	LAC	56	5	1935	66	AA	Howelcke, Johann; Polish astronomer (1611-1687).
Heymans	75.3N	144.1W	50	EU	BE	LMP	3	5	1970	0	AA	Corneille J. F.; Belgian physiologist; Nobel laureate (1892-1968).
Heyrovsky	39.6S	95.3W	16	EU	CZ			5	1985	0	AA	Jaroslav; Czechoslovakian chemist (1890-1967).
Hilbert	17.9S	108.2E	151	EU	GE	LOC	3	5	1970	0	AA	David; German mathematician (1862-1943).
Hill	20.9N	40.8E	16	NA	AM	LTO	043C1	5	1973	0	AA	George William; American astronomer, mathematician (1838-1914).
Hind	7.9S	7.4E	29	EU	GB	LTO	077B3	5	1935	66	AA	John Russell; British astronomer (1823-1895).
Hippalus	24.8S	30.2W	57	EU	GR	LAC	93	5	1935	66	AA	Greek explorer (unkn-c. 120).
Hipparchus	5.1S	5.2E	138	EU	GR	LTO	077B3	5	1935	66	AA	Greek astronomer (unkn-fl. 140 B.C.).
Hippocrates	70.7N	145.9W	60	EU	GR	LMP	3	5	1970	0	AA	Greek doctor (c. 460-377 B.C.).
Hirayama	6.1S	93.5E	132	AS	JA	LTO	082A1	5	1970	0	AA	Kiyotsugu; Japanese astronomer (1874-1943); Shin; Japanese astronomer (1867-1945).

---

lat: latitude of feature center.	**et:** ethnicity of name origin (see page 284 ff.)	**as:** name approval status (see page xvii).
long: longitude of feature center.	**quad:** map quadrangle or informal name (see page xvii ff.).	**ad:** name approval date (year).
diam: diameter or long dimension of feature.		**ref:** reference source for name (see page 287 ff.)
ct: continent of name origin (see page 284 ff.)	**map:** map name or USGS map number (see page xvii ff.)	**ft:** feature type (see page 290).

CRATER	LAT	LONG	KM
ABBE	57.3S	175.2E	67
ABBE H	58.2S	177.9E	25
ABBE K	59.6S	177.3E	28
ABBE M	61.6S	175.5E	29
ABBOT	5.6N	54.8E	10
ABEL	34.6S	85.8E	114
ABEL A	36.6S	86.0E	19
ABEL B	36.7S	82.8E	41
ABEL C	36.0S	81.0E	31
ABEL D	37.7S	87.7E	30
ABEL E	37.8S	86.5E	13
ABEL J	35.5S	79.0E	11
ABEL K	35.0S	77.2E	9
ABEL L	34.4S	82.6E	67
ABEL M	32.2S	83.6E	81
ABENEZRA	21.0S	11.9E	42
ABENEZRA A	22.8S	10.5E	23
ABENEZRA B	20.8S	10.1E	14
ABENEZRA C	21.3S	11.1E	44
ABENEZRA D	21.7S	9.7E	8
ABENEZRA E	21.4S	9.4E	14
ABENEZRA F	21.5S	10.3E	7
ABENEZRA G	20.5S	11.0E	5
ABENEZRA H	21.1S	12.8E	4
ABENEZRA J	19.9S	10.7E	5
ABENEZRA P	19.9S	9.9E	44
ABETTI	19.9N	27.7E	7
ABUL WAFA	1.0N	116.6E	55
ABUL WAFA A	1.4N	116.8E	16
ABUL WAFA Q	0.2N	115.7E	30
ABULFEDA	13.8S	13.9E	65
ABULFEDA A	16.4S	10.8E	14
ABULFEDA B	14.5S	16.4E	15
ABULFEDA BA	14.6S	16.8E	13
ABULFEDA C	12.8S	10.9E	17
ABULFEDA D	13.5S	9.5E	20
ABULFEDA E	16.7S	10.2E	6
ABULFEDA F	16.2S	10.3E	13
ABULFEDA G	13.1S	9.0E	7
ABULFEDA H	13.8S	9.6E	5
ABULFEDA J	15.5S	10.0E	5
ABULFEDA K	14.9S	10.6E	10
ABULFEDA L	14.1S	10.7E	5
ABULFEDA N	16.2S	12.1E	10
ABULFEDA M	15.1S	12.2E	14
ABULFEDA O	15.4S	11.2E	7
ABULFEDA P	15.5S	11.5E	5
ABULFEDA Q	12.8S	12.6E	3
ABULFEDA R	12.8S	13.0E	7
ABULFEDA S	12.2S	13.3E	5
ABULFEDA T	14.8S	13.8E	7
ABULFEDA U	13.0S	13.8E	6
ABULFEDA W	12.5S	13.9E	5
ABULFEDA X	15.0S	14.0E	6
ABULFEDA Y	12.8S	14.1E	5
ABULFEDA Z	14.7S	15.2E	5
ACOSTA	5.6S	60.1E	13
ADAMS	31.9S	68.2E	66
ADAMS B	31.5S	65.6E	28
ADAMS C	32.3S	65.5E	10
ADAMS D	32.5S	71.6E	42
ADAMS M	34.8S	69.2E	24
ADAMS P	35.2S	71.0E	24
AGATHARCHIDES	19.8S	30.9W	49
AGATHARCHIDES A	23.2S	28.4W	16
AGATHARCHIDES B	21.5S	31.6W	7
AGATHARCHIDES C	22.0S	32.9W	12
AGATHARCHIDES E	20.7S	33.0W	15
AGATHARCHIDES F	20.3S	31.8W	6
AGATHARCHIDES G	20.1S	26.7W	6
AGATHARCHIDES H	20.4S	33.9W	15
AGATHARCHIDES J	21.6S	32.5W	13
AGATHARCHIDES K	21.0S	27.4W	11
AGATHARCHIDES L	21.1S	26.7W	8
AGATHARCHIDES N	21.1S	29.6W	22
AGATHARCHIDES O	19.2S	26.6W	5
AGATHARCHIDES P	20.2S	28.7W	66
AGATHARCHIDES R	18.3S	30.7W	5
AGATHARCHIDES S	17.7S	30.5W	3
AGATHARCHIDES T	18.2S	27.7W	5
AGRIPPA	4.1N	10.5E	44
AGRIPPA B	6.2N	9.4E	4
AGRIPPA D	3.8N	6.7E	20
AGRIPPA E	5.2N	8.5E	5
AGRIPPA F	4.4N	11.4E	6
AGRIPPA G	3.9N	6.2E	13
AGRIPPA H	4.8N	10.7E	6
AGRIPPA S	5.3N	8.9E	32
AIRY	18.1S	5.7E	37
AIRY A	17.0S	7.7E	13
AIRY B	17.6S	8.5E	29
AIRY C	19.3S	4.9E	34
AIRY D	18.2S	8.5E	7
AIRY E	20.7S	7.6E	38
AIRY F	18.2S	7.3E	5
AIRY G	18.7S	7.0E	25
AIRY H	18.7S	5.8E	9
AIRY J	19.0S	6.1E	4
AIRY L	20.4S	7.5E	6
AIRY M	19.2S	7.6E	1
AIRY N	17.8S	8.2E	8
AIRY O	16.7S	8.3E	5
AIRY P	15.8S	8.4E	7
AIRY R	19.6S	8.8E	7
AIRY S	17.2S	9.4E	5
AIRY T	19.2S	9.4E	40
AIRY V	17.5S	9.2E	5
AIRY X	18.9S	10.2E	4
AITKEN	16.5S	173.1E	131
AITKEN A	14.0S	173.7E	13
AITKEN C	14.0S	175.8E	74
AITKEN G	16.8S	174.2E	7
AITKEN N	17.7S	172.7E	7
AITKEN Y	12.0S	173.2E	35
AITKEN Z	15.1S	173.3E	33
AL-BAKRI	14.3N	20.2E	12
AL-BIRUNI	17.9N	92.5E	78
AL-BIRUNI C	18.4N	93.0E	9
AL-KHWARIZMI	7.1N	106.4E	65
AL-KHWARIZMI B	9.0N	107.4E	62
AL-KHWARIZMI G	6.9N	107.1E	95
AL-KHWARIZMI H	6.0N	109.2E	50
AL-KHWARIZMI J	6.2N	107.6E	47
AL-KHWARIZMI K	4.6N	107.6E	26
AL-KHWARIZMI L	3.9N	107.4E	35
AL-KHWARIZMI M	3.1N	107.0E	18
AL-KHWARIZMI T	7.0N	104.5E	5
AL-MARRAKUSHI	10.4S	55.8E	8
ALBATEGNIUS	11.2S	4.1E	136
ALBATEGNIUS A	8.9S	3.2E	7
ALBATEGNIUS B	10.0S	4.0E	20
ALBATEGNIUS C	10.3S	3.7E	6
ALBATEGNIUS D	11.3S	7.1E	9
ALBATEGNIUS E	12.9S	6.4E	14
ALBATEGNIUS G	9.4S	1.9E	15
ALBATEGNIUS H	9.7S	5.2E	11
ALBATEGNIUS J	11.1S	6.2E	7
ALBATEGNIUS K	9.9S	2.0E	10
ALBATEGNIUS L	12.1S	6.3E	8
ALBATEGNIUS M	8.9S	4.2E	9
ALBATEGNIUS N	9.8S	4.5E	9
ALBATEGNIUS O	13.2S	4.2E	5
ALBATEGNIUS P	12.9S	4.5E	5
ALBATEGNIUS S	13.3S	6.1E	6
ALBATEGNIUS T	12.6S	6.1E	9
ALDEN	23.7S	110.8E	105
ALDEN B	20.5S	112.6E	17
ALDEN C	22.5S	111.4E	50
ALDEN E	23.2S	112.4E	28
ALDEN V	22.5S	110.1E	19

Fig. 114. (turned opposite) A typical page from the *NASA Catalogue of Lunar Nomenclature*, which contains the lettered craters for all the Moon.

AFTER-WORD

To those readers who have stayed with me to the very end here, I hope that I have conveyed some idea that putting names on lunar features is not just an idle pastime for dilettantes trying to ensure their place in history, but is a serious and essential scientific pursuit akin to naming terrestrial mountains, seas, rivers, etc. Who of us has not been taken aback at some time on reading unfamiliar names of countries such as Burkina-Faso, Nagorno-Kharabakh, Moldova, Myanmar, etc. – not quite what we learned at school! An up-to-date atlas that also gives the previous names is a requisite in order to follow along intelligently.

It's rather similar with the Moon – the latest version of the nomenclature should be available and error-free, and must tie in with the earlier versions if confusion is to be avoided and continuity maintained. One could perhaps envision the predicament of future astronauts aiming to land at the crater 'Hadley' of their LTO charts, only to find that their target is the official 'Hadley', a 15 000-ft mountain peak of Montes Apenninus! 'Houston to Lunar Spacecraft: "Sorry guys; your maps are out of date!"'

APPENDIX A

NAMES IN VAN LANGREN'S MANUSCRIPT MAP

Nomenclature from Van Langren's manuscript map (see fig. 25) compared with that of his engraved map (fig. 26) and modern maps. His use of upper and lower case letters has been retained.

Name on MS map	Name on engraved map	Modern equivalent
Amalfi	same	Barrow
ANNAE	VLADISLAI IV	Tycho
BALTHASARIS	same	Aristarchus
Brahei	same	Aristoteles
Bullialdi	Gassendi	Timocharis
CAROLI	CHRISTIERNI IV	Purbach
CHRISTIERNI	Copernici	Eudoxus
Claramonti	Gutschovii	Dionysius
Deusingii*	Longevilli	Vendelinus
EUGENIAE	same	Plinius
FERDINANDI	FERDINANDI III	Albategnius
Gallilei	Parigi	Riphaeus/Euclides
Gassendi	Haro	Eratosthenes
Golii	Ferd. Francisci	Theophilus
Gutschovii	Puteani	Proclus
INNOCENTII X	same	Ptolemaeus
ISABELLAE	same	Manilius
LADISLAI	CAROLI I	Walter
Lafaillei	Lafaillii	Posidonius
Langreni	same	Langrenus
Lantsbergii	Noyelles	Piccolomini

Name on MS map	Name on engraved map	Modern equivalent
Leurechonii	Mexiae	Reinhold
LUDOVICI XIV	same	Alphonsus
MARIAE	same	Menelaus
Medices	Medicaei	Bullialdus
Moura	same	Cleomedes
PHILIPPI	PHILIPPI IV	Copernicus
Raeithaei**	Bazan	Pytheas
Ricardii	Pozzo	Aristillus
Sempili	Kintschotii	Lansberg
S. Marci	same	Gassendi A
Tacquetti	same	Prom. Laplace alpha
Vendelini	Wendelini	Maskelyne A
Fretum Catholicum	same	S. Aestuum/M. Vaporum
Mare Astronomorum	MARE ASTRONOMICUM	M. Frigoris
MARE BELGICUM	same	M. Tranquillitatis
MARE EUGENIANUM	same	M. Serenitatis
MARE LANGRENIANUM	same	M. Fecunditatis
MARE DE MOURA	MARE DE MOURA sive CASPIUM	M. Crisium
MARE VENETUM	same	M. Humorum
Montes Austriaci	same	Montes Apenninus
OCEANUS PHILIPPICUS	same	Oceanus Procellarum
Prom. Clavi	Prom. Clavii	Prom. Laplace
Prom. S. Vincenti	Prom. S. Vincetii	Prom. Heraclides
Sinus Austriacus	MARE AUSTRIACUM	M. Imbrium
SINUS BATAVICUS	same	N. Nectaris+ S. Asperitatis
Sinus Mediceum	MARE BORBONICUM	M. Nubium
SINUS PRINCIPIS	same	S. Roris

* Name not used on engraved map.
** Rheita.

APPENDIX B

DIFFERENCES BETWEEN VAN LANGREN'S ENGRAVED MAPS

These are the differences between the three states of Van Langren's engraved map.

	State 1 (Leyden)	State 2 (Edinburgh)	State 3 (Paris and San Fernando)
author's name	VAN LANGREN	LANGRENUS	LANGRENUS
Sabine crater	Cambieri	Cambierii	Cambierii
Petavius crater	Tristis	Tristis	Triestis
Nasireddin crater	Conde	Prin. Conde	Prin. Conde

APPENDIX C

DIFFERENCES BETWEEN THE STRASBOURG FORGERY AND PARIS VERSION

These are the differences between the Strasbourg forged map and the engraved maps.

Strasbourg	Third state	Modern
Barancii	Bazan	Pytheas
Campinii	Haro	Eratosthenes
Monconisii	Claramontii	Bianchini
Neuraei	Morini	Harpalus
Prom. S. Pui	Prom. S. Petri	anon.
Silvecani	Albategni	Euctemon?

Missing names: Arati, Portus Gallicus, Prom. S. Ignatii, Sinus Opticus, Rho.

APPENDIX D

....

VAN LANGREN'S NOMENCLATURE

Nomenclature from Van Langren's engraved map of 1645 (see fig. 26). Listed alphabetically in sections according to his original print type, which almost completely separates the names of philosophers, astronomers, etc. from those of the nobility, royalty, etc.

A *NAMES IN ITALICS*; CRATERS, ISOLATED PEAKS, LIGHT OR DARK SPOTS

Name on map	Modern equivalent	Name on map	Modern equivalent
*Albategni	Euctemon?	*Briggi	Curtius B
Andradae	M. Crisium omega	*Bullialdi	Reiner gamma
*Arati	Plato tau	Caleni	Asclepi
*Archimedis	Teneriffe iota	Cambierii	Sabine
*Aristarchi	Archytas	*Caramuelis	Descartes brt. area
Auberi	Nicolai	*Cartesii	Römer
Badvari	anon. bright area	Chisletti	Lambert
*Baieri	Delisle	Ciermanni	Brayley
Bakii	Arago	*Claramontii	Bianchini
Barlaei	Atlas A or alpha	Cobavi	Pico
Barreae	Burckhardt	Cocci	Vitruvius
Bechleri	Triesnecker	Coci	Teneriffe delta etc.
Bervoeti	Ramsden	Conradi	Wrottesley
*Bettinii	Bessarion & A	Contarini	anon. brt. area
Biaei	Hercules	Conti	T. Mayor alpha
Bickeri	Vitello	*Copernici	Eudoxus
*Blancani	Flamsteed rho etc.	Cornaro	anon. brt. area
Blitterswyckii	Atlas	*Crugeri	Prom. Heraclides ?
Bonvicini	Riccioli D dk. spot	*Curtii	Furnerius?
*Brahei (=Tycho)	Aristoteles	*Cusae	Eudoxus kappa

Name on map	Modern equivalent	Name on map	Modern equivalent
D'Auxoni	Montes Recti beta	Leototi	Berosus
Danesii	Clairaut	Leurechonii	Spitzbergen
*Derienni	Ideler & L	*Longomontani	Mason or Plana?
Derkenni	Abulfeda	Lutiani	Grove
Edelherii	Geminus	*Magini	Liebig gamma
*Endymionis	Endymion	Magni	Capella B?
*Euclidis	Luther epsilon	Marci	T. Mayer A
*Eychstadi	Werner	*Mersenni	Azophi
*Finiae	La Hire	*Moleri	Baco B?
*Fournerii	Darney	*Moreti	Picard gamma
Fromi	Blanchinus	Morgues	dk. spot nr. Fourier
Fromondi	Ross	*Morini	Harpalus
*Gallilaei	Campanus	Nachara	=Aytona
Gansii	Halley	Naudei	Argelander
Garsioli	Kepler theta	Navei	Maskelyne
*Gassendi	Timocharis	*Neperi	Lubbock N
Gevartii	Rabbi Levi	Nirenbergeri	Wurzelbauer D
Ginnari	Brayley B	Nobelarii	Drebbel E dk. spot
Giovanelli	Mersenius zeta	Nuti	Zach
Golii	Censorinus N	Pappi	M. Humboldtianum
Grassi	Hahn	Parigi	Riphaeus brt. area
*Grimbergeri	Fra Mauro sigma etc.	Phorylidi	Mädler
Gualteri	Hortensius?	Piperii	=Silgero
Guldini	Palmieri alpha	Pironi	Kant
Gutschovii	Dionysius	*Pitati	Pico beta
Haesteni	Mairan	Pozzo	Aristillus
Hardii	Parry	Pratii	Bode?
Hensii	Franklin	*Ptolomaei	C. Mayer
Herlici	Hansteen alpha	Puteani	Proclus
*Hevelii	Lubiniezky A	*Pythagorae	Pythagoras
Hugenii	Mercurius	*Pythias	Hercules A
*Hypatiae	Sulp. Gallus gamma	Quaresini	Lee eta
*Hypparchi	Scoresby	Rechbergeri	Sharp
*Kepleri	Protagoras zeta	*Regiomontani	Endymion C
*Kircheri	Malapert	*Reithae	Autolycus
Lafaillii	Posidonius	Rho	Clairaut A?
*Langreni	Langrenus	*Ricci	Euler beta
*Lantsbergi	Breislak?	Richardi	Marius B
Laucii	Kepler gamma	Robervalis	Wolf
Le Pessier	Euler	Rubenii	Lacus Timoris (NC)

Name on map	Modern equivalent	Name on map	Modern equivalent
*S. Bedae	Langrenus B+K+F	Trederi	Zöllner
Scala	Cruger dark spot	Tucheri	Macrobius
*Scheineri	Louville	Valerii	Playfair
*Schonbergeri	Piton	Vici	dk. spot in Schickard
Schotenii	R. Recti epsilon?	*Vlacci	Capella
*Schyrlei	Abenezra	Vossii	Cepheus
Scialli	Kunowsky	Vulleri	Alhazen alpha
Seneschali	Alfraganus alpha	Wassenarii	Torricelli
*Simpilii	Marius A	Wegii	Pentland
*Snellii	Isidorus F?	Welperi	Colombo? (not marked)
Stratii	Ukert	*Wendelini	Maskelyne A
*Tacquetti	Prom. Laplace alpha	Wisilii	Street
*Thales	Bürg	*Wolfii	Lindenau
*Thebit	Luther zeta	*Xenophanis	Strabo
*Timochari	Timaeus	Zylii	Theaetetus
Tirelli	Diophantus	Annulus Neptuni	Gassendi
Torii	M. Crisium tau?		

B *NAMES IN ITALICS*; OTHER FEATURES

Name on map	Modern equivalent	Name on map	Modern equivalent
Prom. Argolii	anon.	Sinus *Erathostenis	anon.
Prom. *Arzahel	anon.	Lacus Masii	Lacus Excellentiae (NC)
Prom. Calippi	anon.	Lacus *Possidoni	Grimaldi dk. area
Prom. Cassiodori	anon.	Portus Adriaticus	anon.
Prom. Cesaris	anon.	Portus Gallicus	anon.
Prom. *Clavii	Prom. Laplace	Portus S. Francisci	Pitatus+Hesiodus dk. area
Prom. *Cleomedis	anon.	Aestuaria Bamelrodia	Palus Somni
Prom. *Methonis	anon.	Regius Fluvius	M. Spumans+M. Undarum
Prom. *Procli	anon.	Flumen S. Augustini	anon.
Mons S. Xaverii	E. wall of Cassini		

C *NAMES IN CAPITAL ROMAN LETTERS*; CRATERS, ISOLATED PEAKS, LIGHT
& DARK SPOTS. ALL NAMES DROPPED FROM SUBSEQUENT MAPS.

Name on map	Modern equivalent	Name on map	Modern equivalent
ANNAE Reg. Fran.	Arzachel	CAROLI I Reg. Britt.	Walter
BALTHASARIS		CHRISTIERNI IV	Purbach
Hispa. Pri.	Aristarchus	Reg. Daniae	

Name on map	Modern equivalent	Name on map	Modern equivalent
CHRISTINAE Reg. Suec.	Regiomontanus	LUDOVICI XIV Reg. Fran.	Alphonsus
EUGENIAE	Plinius	MARIAE Imp. Rom.	Menelaus
FERDINANDI III Imp. Rom.	Albategnius	PHILIPPI IV	Copernicus
		ROMA	Archimedes
INNOCENTII X	Ptolemaeus	VLADISLAI IV Reg. Pol.	Tycho
ISABELLAE Reg. Hisp.	Manilius		

D *NAMES IN ROMAN CAPITALS*; OTHER FEATURES. ALL NAMES DROPPED FROM SUBSEQUENT MAPS.

Name on map	Modern equivalent	Name on map	Modern equivalent
TERRA DIGNITATIS	anon.	MARE ASTRONOMICUM	M. Frigoris
TERRA HONORIS	anon.	MARE AUSTRIACUM	M. Imbrium
TERRA IUSTITIAE	anon.	MARE BELGICUM	M. Tranquillitatis
TERRA LABORIS	anon.	MARE BORBONICUM	N. Nubium
TERRA PACIS	anon.	†MARE CASPIUM (DE MOURA)	M. Crisium
TERRA SAPIENTIAE	anon.	MARE EUGENIANUM	M. Serenitatis
TERRA TEMPERANTIAE	anon.	MARE LANGRENIANUM	M. Fecunditatis
TERRA VIRTUTIS	anon.	MARE VENETUM	M. Humorum
LITTUS PHILIPPICUM	anon.	FRETUM PACIS	anon.
OCEANUS PHILIPPICUS	Oceanus Procellarum		

E *NAMES IN ROMAN LOWER CASE LETTERS*; CRATERS, ISOLATED PEAKS, LIGHT AND DARK SPOTS

Name on map	Modern equivalent	Name on map	Modern equivalent
*Alfonsi IX Reg. Cast.	Democritus	Benavidi	Gay-Lussac nu?
Amalfi	Barrow	Bichi	Riccius
Annae D. Aurel. F.	Caucasus gamma	Boivinii	Newcomb
Anselmi Elect. Mogunt.	Casatus	Bracamonti	Polybius
		Brunii	Macrobius A
Arenbergii	Murchison+Pallas		
Arondelii	Lexell A	Cantelmi	Cichus
Aytona	Harbinger beta	Caroli D. Loth.	Metius
Barbarini	Mösting	Caroli D. Mant	Jacobi
Bazan	Pytheas	Casimiri Pol. P.	Mutus
Bekii	Godin	Cerda	Krieger

Name on map	Modern equivalent	Name on map	Modern equivalent
Chigi	Tacitus	Mariae D. Mont. F.	Hadley
Clarae Isab. Leop. F.	Sulp. Gallus M brt. spot	Mariannae Imper. F.	Calippus alpha
Conde Prin.	Nasireddin	Martinitzi	Bettinus
Crani	Delambre	Masii	Gauricus
Crequii	Airy	Maximiliani Duc. Bava.	Clavius
Croii	Licetus+Heraclitus	Mazarinii	Alpetragius
Cuevio	Harbinger alpha	Medicaei	Bullialdus
Doriae	Hell	Mexiae	Reinhold
Elisabethae Palat. Fil.	Aliacensis	Moura	Cleomedes
		Nassauii	Fracastorius
Emanuelis D. Sab.	Stiborius		
Estensis D. Mutinae	Maurolycus	Noyelles	Piccolomini
Farnesii D. Parmae	Orontius	Ocariz	Julius Caesar
Ferd. Caroli Leop. F.	Cyrillus	Ossolinski	Sirsalis A brt. spot
Ferd. Francisci Imp. Rom. F.	Theophilus	Oxensterni	Snellius
		Pamphilii	Herschel
Ferdinandi Elect. Col.	Blancanus		
Francisci D. Loth.	Kircher	Philip Christ. Elect. Treu.	Vlacq
Fred. C. Pal.	Fabricius		
Fred. Wilhelmi M. Brand.	Schiller	*Piccolomini	Catharina
		Quesada	Milichius gamma
Frederici D. Holsat.	Newton	Radsevillii	Gemma Frisius
Gastoni D. Aurel.	Airy B	Ramirii	Firmicus
Gauraei	Santbech	Rantsovii	Thebit
Guasco	Pitiscus	Recki	Hainzel
Haro	Eratosthenes	Rosetti	Lalande
Ioanni D. Sax.	Longomontanus	S. Marci	Gassendi A
Isenburgi	Almanon	Saavedrae	dk. spot in Schickard
Kintschotii	Lansberg	Segueri	Stevinus
Konie c Polski	Hommel	Sfondrati	Seleucus
Laurini	Isidorus	Silgero	Harbinger delta+eta
Lennoxis	Apianus	Slavatae	Horrocks
Leopoldi Arch. Aust.	Maginus	Spada	Marius
Ligni	Furnerius?	Spinola	Lilius
Lini	Teneriffe epsilon	Tassis	Geber
Longevalli	Zagut	Taye	Boscovich
Longevilli	Vendelinus	Theresae Hispa. Inf.	Calippus theta+eta+omega
*Malvezzi (=Malvasia?)	Byrgius A brt. spot	Thomae D. Sab.	Kepler

Name on map	Modern equivalent	Name on map	Modern equivalent
Trautmansdorffii	Hind	*Wilhelmi	Mee
Triestis	Petavius	Lantgravii	
*Ulloae	Taylor	Wolfgangi D.	Wilhelm
Urselii	Agrippa	Neoburgi	
		Zamosci	Watt+Steinheil

F *NAMES IN ROMAN LOWER CASE LETTERS;* OTHER FEATURES

Name on map	Modern equivalent	Name on map	Modern equivalent
Montes Austriaci	Montes Apenninus	Prom. Salamona	anon.
Prom. *Colombi	anon.	Mare de Popoli	Palus Epidemiarum
Prom. Henrizi D.	anon.	Sinus Athlanticus	anon.
Venet.		Sinus Batavicus	M. Nectaris+S. Asperitatis
Prom. S. Alberti	anon.		(NC)
Prom. *S. Dionisii	anon.	Sinus Geometricus	S. Iridum
Prom. S. Dominici	anon.	Sinus Geometricus	S. Iridum
Prom. S. Iacobi	anon.	Sinus Medius	S. Medii
Prom. S. Ignatii	anon.	Sinus Opticus	S. Amoris (NC)
Prom. S. Ludovici	anon.	Sinus Principis	S. Roris
Prom. S. Michaelis	anon.	Fretum Catholicum	S. Aestuum/M. Vaporum
Prom. S. Petri	anon.	Lacus Panciroli	Plato
Prom. S. Vincetii	Prom. Heraclides	Lacus Scientiae	L. Somniorum, (pE. part)

Names in the second column are from *Named Lunar Formations*, IAU 1935, except where superseded by *The System of Lunar Craters*, 1963–6, and the *NASA Catalogue of Lunar Craters*, 1982 (labelled 'NC'). Subsequent changes (1985–95) do not affect any of the listed names or designations.

† Mare Caspium used 2 years later by Hevelius, but on different mare.

* Names used by Riccioli and/or later mappers (often with slightly changed spelling), but not necessarily appearing on modern maps.

HEVELIUS'S NOMENCLATURE

Nomenclature from Hevelius's list and map, listed alphabetically by name and not by category of feature type. A comma indicates that the order is reversed on the map, thus **M. Abarim**, but **Aconitus Collis**.

Name from list or map	*Modern equivalent*
Abarim, mons	crater chain, Pitatus-Ball
Acabe, mons	Byrgius+brt. nimbus
Aconitus collis	Boscovich
Adriaticum, mare	Sinus Medii+SE. Sinus Aestuum
Aea, insula	Cauchy
Aegyptiacum, mare	part of S. Oc. Procellarum
Aegyptus (regio)	SW highlands
Aemus (or Haemus), mons	Alexander
Aerii, montes	M. Carpatus, W. end
Aethusa, insula	brt. spot SW. of Lansberg D
Aetna, mons	Copernicus
Africae pars	far W. highlands
Agarum, prom.	Prom. Agarum
Ajax, mons	Cavendish
Alabastrinus, mons	Lichtenberg brt. nimbus
Alani, montes	N. rim mts. of M. Crisium
Alaunus, mons	Plutarch+Seneca
Alopecia, insula	Picard
Alpes (montes)	Montes Alpes
Amadoca, palus	M. Humboldtianum
*Amadoci, montes	Messala area

201

Name from list or map	Modern equivalent
Amanus, mons	Delambre
Amarae, paludes	M. Spumans+M. Undarum
Amari, fontes	Crüger dk. floor
Ambenus, mons	brt. ray, Mason-M. Frigoris
**Anemusa, insula	prob. brt. spot N. of Lansberg
Annae, mons	Wilhelm+Longomontanus
Antilibanus, mons	Werner+Aliacensis
Antitaurus, mons	Playfair to Almanon (chain)
Apenninus, mons	Montes Apenninus
**Apollinis, prom.	prob. Prom. Laplace
*Apollinis, sinus	Sinus Iridum
Apollonia, insula	Plinius
Apollonia Minor, insula	Ross
Arabia (regio)	SSW limb area
Arabiae, paludes	dk. area De Gasparis-Palmieri
Archerusia, palus	Julius Caesar
Archerusia, prom.	Prom. Archerusia
Areesa, palus	dk. area SW. of Fracastorius
Argentarius, mons	Archimedes
Arietis, prom.	near Vitruvius
Armeniae, montes	Gemma Frisius+Goodacre
Arrhentias, insula	brt. area N. of Moltke
**Asiae pars	prob. SSE limb area
*Asia Minor (regio)	central highlands
Atheniensis, sinus	Sinus Asperitatis
Athos, mons	Euclides brt. nimbus
Atlas Major, mons	Bouguer+La Condamine
Atlas Minor, mons	Sharp+Sharp B
Audus, mons	Reiner gamma (+Reiner?)
Aureus, mons	Egede A brt. nimbus
Baronisus, mons	Mairan
Besbicus, insula	Manilius
Berosus, mons	Römer
Bodinus, mons	Cepheus+Franklin
Bontas, mons	Democritus
Borysthenes, lacus	L. Somniorum, W. portion
**Bosphorus, sinus Propont. ad	M. Vaporum, NE corner
Byces, palus	Sinus Amoris
Byzantium (urbs)	Menelaus

Name from list or map	Modern equivalent
Cadmus, mons	Parrot C
**Calabraria, insulae	prob. in M. Teneriffe
Calathe, insula	Tob. Mayer beta, rho etc.
Calchistan, mons	Stöfler+Maurolycus etc.
Capraria, insula	M. Teneriffe epsilon
Carcinites, sinus	L. Somniorum, E. portion
Carpates, mons	Eudoxus
Carpathos, insula	Max Wolf
Casius, mons (#1)	Vieta+Fourier
*Casius, mons (#2)	Agatharchides P
Caspium, mare	M. Fecunditatis
Cassiotis regio	E. border of M. Humorum
Cataractes, mons	Gassendi
Caucasius, sinus	Sinus Concordiae
Caucasus, montes	M. Pyrenaeus
*Caucasus Inferior, montes	Santbech+Borda etc.
Celenorum tumulus	La Caille to Vogel (chain)
Cercinna, insula	Kepler lge. brt. nimbus
Chadisia, prom.	near Moltke
Chalcidici, montes	ray E. from Copernicus
*Christi, mons	Piton
Cilicum, insula	Arago
Cimaeus, mons	Halley+Hind
**Cimmeriae, paludes	L. Bonitatis etc.
Cimmerius, mons	Macrobius
Circaeum, prom.	near Timocharis
**Cirna, mons	unknown
Climax, mons	Sirsalis+A+brt. nimbus
Coibacarani, montes	Fabricius+Metius+Brenner
Colchis (regio)	E. of S. Asperitatis
Corax, mons	Proclus+SW of rim M. Crisium
Corocondametis, lacus	P. Somni
Corsica, insula	Timocharis
Cosyra, insula	Tob. Mayer A brt. nimbus
Cragus, mons	Arzachel
Cratas, mons	ray N. from Copernicus
Crathis, mons	Fra Mauro kappa, gamma, zeta
Creta, insula	Bullialdus
**Creticum, mare	prob. near Bullialdus)
Cyanea Europa, insula	vague brt. area N. of Tacquet

Name from list or map	*Modern equivalent*
Cydises, mons	Apianus
Cyprus, insula	Birt
Delanguer, mons	Vlacq+Hommel+Nearch etc.
Didymae, insulae	Ramsden+Capuanus
Didymus, mons	Albategnius
Ebissus, insula	Pico beta
Echinades, insulae	brt. spots W. of Fra Mauro A
Eos, mons	brt. & dk. areas E. of Piazzi
Eoum, mare	W. part Oc. Procellarum
Erichtini, scopuli	Jansen and Arago brt. areas
Erroris, insula	Helicon+Leverrier
Eryx, mons	rays NNE from Copernicus
Evila, desertum	Deluc-Gruemberger area
Extremus Ponti, sinus	M. Nectaris
Ficaria, insula	Euler
Fortis, mons	Nonius+Kaiser
**Freti Pontici, prom.	brt. point at Madler
Gallicus, sinus	NE. part of M. Imbrium
Germanicianus, mons	Marius
Hajalon, vallis	Scheiner+Blancanus
Hellespontum, sinus Propont. ad	W. corner of M. Vaporum
Heracleum, prom.	brt. point at Censorinus
Herculeum, prom.	brt. point at Da Vinci
Herculeus, lacus	3 dk. spots SE of Copernicus
Herculis, mons	Maskleyne A+Censorinus A nimbus
Hereus, mons	ray NNW from Copernicus
Hermo, mons	ray NNE from Tycho
Hiblaei, montes	ray WSW from Copernicus
Hiera, insula	Pytheas
Hippoci, montes	Kapteyn+Maclaurin F
Hippolai, prom.	brt. point near Luther
Hipponiates, sinus	Sinus Aestuum
Hor, mons	Heinsius P, Q, etc.
Horeb, mons	Heinsius+A, B, and C
Horminius, mons	Dionysius+D'Arrest etc.
Hyperborea, regio	N. polar limb area
Hyperboreae, paludes	vague dk. area N. of Berzelius
Hyperborei, montes	Fontenelle A to Scoresby
Hyperborei, scopuli	Protagoras+Archytas
Hyperboreum, mare	M. Frigoris

Name from list or map	Modern equivalent
*Hyperboreus, sinus	Sinus roris
Hyperboreus Inferior, lacus	L. Spei
Hyperboreus Superior, lacus	Endymion
Ida, mons	Godin+Agrippa
Inferior, sinus	S. end of M. Fecunditatis
Inferiores, paludes	ray NNE from Bessel
**Insula	Dawes+nimbus
Italia (regio)	brt. areas W. of M. Apenninus
Lathmus, mons	Palisa+Davy
Lea, insula	brt. spot at Letronne pi (mts.)
Lemnos, insula	Lalande A+brt. area
Lesbos, insula	Lassell+B, C etc.
Letoa, insula	Campanus A nimbus
Leucopetra, prom.	Prom. Agassiz
Libanus, mons	Purbach+Regiomontanus+Walter
Ligustinus, mons	Aristillus
Lion, mons	Hainzel+Mee etc.
Lipulus, mons	Zollner
Lunae, prom.	vague brt. area at Spitzbergen
Lychnitis, palus	dk. area SW of Santbech
Lybiae pars	WSW limb area
Macra, insula	Posidonius
Macrocemnii, montes	Atlas+Hercules
Maeotis, palus	M. Crisium
Major, Insula	Langrenus
Major Occidentalis, lacus	M. Marginis
Majorca, insula	Montes Recti
Malta, insula	Lansberg
Mampsarus, mons	brt. area at Seuss
Mantiana, palus	dk. area SE of Fracastorius
Maraeotis, palus	Grimaldi
Marinus, lacus	W. corner of M. Imbrium
Masicytus, mons	Alphonsus
Mauritania (regio)	anon. (E. and N. of Sns. Iridum)
Mediterraneum, mare	M. Imbrium, Nubium+E. Oc. Pr.
Melos, insula	SE border of M. Cognitum
Menyx, insula	brt. area at Kunowsky
**Mercurii, prom.	unknown, but prob. nr. Aristarchus
Meridionalis, lacus	dk. area SW of Schiller
Meridionalis, mons	Kircher+Bettinus

Name from list or map	Modern equivalent
Mesogys, mons	Herschel
Micale, mons	Lalande
**Mimas, mons	prob. Mosting
Mingui, desertum	S. limb area
Minor, insula	Naonobu+Bilharz+Atwood
Minor Occidentalis, lacus	dk. areas Hansen B and to W.
Minorca, insula	Pico
Miris, stagnum	Riccioli O dk. spot
Moesia (regio)	N. border of M. Serenitatis
Montuniates, mons	Autolycus
Mortuum, mare	Pitatus dk. floor
*Moschus, mons	Theophilus+Cyrillus+Catharina
*Mundi, catena	NW border mts. of M. Serenitatis
Myconius, mons	ray NE from Copernicus
Mysius, mons	Rhaeticus+Dembowski
Neptunus, mons	Reinhold
Nerossus, mons	Petavius+Hase+Furnerius
Niger Major, lacus	Plato
Niger Minor, lacus	dk. spot N. of Alpes
Nilus, fluvius	dk. streaks at Billy – Mersenius
Nitriae, mons	dk. spots at Lepaute D
**Occidentalior, insula	unknown
Olympus, mons	Hipparchus
Ophiusa, insula	M. Teneriffe delta etc.
*Orient, paludes	NW edge of Oc. Procellarum
**Orientalior, insula	unknown
**Orientalis, sinus	W. end of M. Frigoris
Paestanus, sinus	SSE area of M. Imbrium
*Palaestina (regio)	SSW highlands
**Palaestinae, paludes	unknown
Paludes	ESE limb area
Paludosa, loca	inner nimbus of Kepler
Pamphylium, mare	S. part of M. Nubium
Pangaeus, mons	Pallas+Murchison+Ukert
Paropamisus, mons	Snellius A+Furnerius A nimbi etc.
Parthenius, mons	Fra Mauro zeta
Peloponnesus (peninsula)	brt. area, Guericke to Gambart
Pentadactylus, mons	Seleucus
Peronticus, sinus	NW corner of M. Serenitatis
Persia (regio)	SSE highlands

Name from list or map	Modern equivalent
Peuce, mons	NW border of L. Mortis
Pharan, montana	Clausius region
Phasianus, sinus	ESE corner T. Tranquillitatis
Pherme, mons	Hevelius
Philyra, insula	Maskelyne
Phoenix, mons	Alpetragius
Pontia, insula	Wichmann+mts. gamma, theta etc.
Ponticum, fretum	N. part of M. Nectaris
Pontus Euxinus (mare)	M. Serenitatis+M. Tranquillitatis
Porphyrites, mons	Aristarchus
Propontis (sinus)	S. part of M. Vaporum
Prophetarum, mons	Wurzelbauer D
Pyramidalis, petra	Mts. Spitzbergen
Raphidim, desertum	W. part of Clavius
Rhodus, insula	brt. area at Nicollet+delta etc.
Riphaei, montes	Cleomedes, Burckhardt, Geminus etc.
*Romania (regio)	SE of M. Apenninus
**Rupes	unknown
Sacer, mons	W. border of M. Humorum
Sagaricus, sinus	NNE part of M. Serenitatis
St. Petro, insula	Mons Vinogradov
Salmydessus, sinus	brt. patch at Linne
Sanctus, mons	Apollonius+Firmicus
Sardinia, insula	Lambert
Sarmatiae Asiaticae pars	S. of M. Crisium
Sarmatiae Europaeae pars	ENE limb area
Sarmatici, montes	Chr. Meyer
Scithiae pars	SW limb area
Seir, montana	Street+Maginus etc.
Sepher, mons	ray NW from Tycho
Serrorum, mons	Aristoteles
Sicilia (insula)	Copernicus nimbus
Sinai, mons	Tycho
Sinopium, mons	dk. area at Elger E
Sipylus, mons	Ptolemaeus
Sirbonicum, fretum	Oc.Proc. – M. Humorum join
Sirbonis, sinus	M. Humorum
Sogdiana, petra	dk. area at S. end of Vendelinus
Sogdiani, montes	Piccolomini
Strobilus, mons	Isidorus (+Capella?)

Name from list or map	Modern equivalent
Strophades, insulae	Darney chi, tau, lambda (mts.)
Strymonicus, sinus	E. part of M. Nubium
Superiores, paludes	faint brt. line, W. M. Serenitatis
Syrticum, mare	SW corner in O. Procellarum
Syrticus, sinus	dk. area SW of Encke
Tabor, mons	very bright nimbus at Deslandres QA
*Tadnos, fons	two dk. floor patches in Schickard
Taigetus, mons	Guericke
Tancon, mons	Colombo
Tarantinus, sinus	SE part of M. Insularum
Taurica chersonnesus	Montes Taurus+area to S.
Taurus, mons	brt. ray E. from Tycho
Taraciniae, insulae	Hortensius nimbus+brt. spot to W.
Techisandam, mons	Moretus+Curtius
Thambes, mons	vague brt. area NE of Hermann
Taenarium, prom.	brt. point at Guericke B
Thospitis, lacus	Fracastorius
Tmolus, mons	Taylor+A etc.
Tornese, caput de	brt. point at Fra Mauro B)
Trapezus, mons	Newcomb
Trasimenus, lacus	P. Putredinis
Troicus, mons	NE wall of Schickard
Uxii, montes	Zagut, Rabbi Levi, Lindenau etc.
Vulcania, insula	Eratosthenes
Zacynthus, insula	brt. area E. of Lansberg
Zin, desertum	N. half of dk. halo around Tycho

* Name on map, but not in list (14 in all).
** Name in list, but not on map (17 in all). General locations or actual identifications are possible in most cases from other data given in the list.
Modern names are from the same sources as for Appendix D.

APPENDIX F

HEVELIUS'S NAMES STILL USED IN MODERN MAPS

Hevelius's names in use on modern maps, a) in same location, b) moved

a) Agarum, Prom.
  Alpes, (Montes)
  Apenninus, Montes
  Archerusia, Prom.

b) Carpates, Mons
  Caucasus, Montes
  Haemus, Mons
  Riphaei, Montes
  Taurus, Mons
  Taenarium, Prom.

APPENDIX G

....

RICCIOLI'S NOMENCLATURE

Nomenclature from Riccioli's list and map, listed by feature type.

A CRATERS; IRREGULAR ENCLOSURES BETWEEN MOUNTAINS ETC.; SPOTS OR SMALL AREAS, LIGHT OR DARK.

Name from list or map	Modern spelling	Modern feature name
Abenezra		same
Abilfedea	Abulfeda	same
Agrippa		same
Albategnius		same
*Alcuinus		Lubbock N
Alfraganus		Zöllner
Aliacensis		same
Almaeon	Almanon	same
Alpetragius		same
Alphonsus Rex	'Rex' dropped	same
Anaxagoras		Goldschmidt
Anaximander		Carpenter
Anaximenes		Pascal
Apianus		same
Aratus		'bay' S. of M. Hadlery
Archimedes		same
Architas	Archytas	Egede A+nimbus
Ariadaeus		Dionysius
Aristarchus		same
Aristillus		same

Name from list or map	Modern spelling	Modern feature name
Aristoteles		same
Arzachel		same
*Arzet	(in list, not map)	Zach (probably)
Atlas		same
Autolicus	Autolycus	same
Azophi		same
Barocius		same
*Bartolus		Bailly B
Bayerus	Bayer	Schiller
*Beda		Censorinus N
Berosus		Hahn
Bessarion		same
Bettinus Soc. I.	'Soc. I.' dropped	same
Billy		same
Blancanus		same
Blanchinus		same
Bullialdus		same
Byrgius		same
Cabeus		Newton
Calippus		Alexander
Campanus		same
Mart. Capella	'Mart.' dropped	same
Capuanus		Ramsden
Cardanus		same
Casatus	(not in list)	same
St. Catharina	'St.' dropped	same
Cavalerius		same
Censorinus	(not in list)	same
Cepheus		Franklin
Cichus		Capuanus
*Claramontius	(not in list)	(darkish area)
Clavius Soc. I.	'Soc. I.' dropped	same
Cleomedes		same
Cleostratus		same
Conon		'bay' SW of M. Bradley
Copernicus		same
Crugerus	Cruger	same
Curtius Soc. I.	'Soc. I.' dropped	same
Cusanus		T. Mayer alpha
S. Cyrillus Alex.	Cyrillus	same

Name from list or map	Modern spelling	Modern feature name
Cysatus Soc. I.	'Soc. I.' dropped	same
*Dantes		Gambart C
Democritus		same
*Deriennes		Letronne A brt. spot
*Dionysius Exiguus		Censorinus C
S. Dionysius	'S.' dropped	Delambre
*Dominicus Maria		dark area
*Ecphantus		Gruit. gamma+delta
Eichstadius	Eichstadt	same
Endymion		same
Epigenes		same
Eratosthenes		same
Euctemon		same
Eudoxus		same
*Eustachius		Siralis Z
Fabritius	Fabricius	same
Fernelius		Nonius
Firmicus		same
Fontana		Zupus dark spot
Fracastorius		same
Furnerius Soc. I.	'Soc. I.' dropped	same
Galilaeus	Galilaei	Reiner gamma
Gassendus	Gassendi	same
Gauricus		same
Geber		same
Geminus		same
Gemma Frisius		same
Goclenius		same
Grimaldus Soc. I.	Grimaldi	same
Gruemberger Soc. I.	'Soc. I.' dropped	same
Gulielmus Hassiae	Wilhelm	same
Hagecius		Hommel
Hainzelius	Hainzel	Mee
Harpalus		same
Helicon Cyzicenus	'Cyzicenus' dropped	same
Heraclides Ponticus	Prom. Heraclides	same
Hercules		same
Herigonius		Herig. delta brt. area
*Hermes		dark area
Hevelius		same

Name from list or map	Modern spelling	Modern feature name
Hipparchus		same
Homelius	Hommel	Pitiscus
Hortensius		same
Higinus	Hyginus	dark spot
Hypatia		same
S. Isidorus Hisp.	Isidorus	same
Iulius Caesar	Julius Caesar	Boscovich
*Iunctinus		Fra Mauro sigma
Keplerus	Kepler	same
Kircher Soc. I.	'Soc. I.' dropped	same
*Kristmannus		L. Excellentiae
Langrenus		same
Lansbergius	Lansberg	same
Licetus		Cuvier
Lilii fratres	Lilius	Lilius, Jacobi+A
*Linemannus		Flamsteed kappa etc.
Longomontanus		same
Macrobius		same
Maginus		same
Manilius		same
Malapertius	Malapert	Simpelius B
Manzinus		Boguslawsky
Simon Marius	'Simon' dropped	same
Maurolycus		same
Menelaus		same
Metius		Brenner
Meton		same
Mercurius		L. Spei
Mersenius		same
Messala Arabs.	'Arabs.' droppped	same
Milichius		Milichius alpha
*Moletius		Fra Mauro lambda
Moretus Soc. I.	'Soc. I.' dropped	same
*Morinus		bright area
*Mulerius		Metius
*Munosius		Wolf
Mutus		Boussingault B+E
Neander		same
Nonius		Kaiser
Oenopides		same

Name from list or map	Modern spelling	Modern feature name
*Origanus		Darney tau+chi brt. spot
Orontius		same
*Osymandies		Mercurius
Petavius Soc. I.	'Soc. I.' dropped	same
Philolaus		same
Phocylides		same+Nasmyth
Piccolomineus	Piccolomini	same
Pitatus		same
Pitheas Massil.	Pytheas	same
Pitiscus		Vlacq
Plato		same
Plinius		same
Plutarchus	Plutarch	same
Pontanus		same
Possidonius	Posidonius	same
Proclus		same
*Profatius		Nicollet B brt. spot
Ptolemaeus		same
Purbachius	Purbach	same
Pythagoras		same
Rabbi Levi		same
Regiomontanus		same
Reinerus	Reiner	same
Reinholdus	Reinhold	same
Reitha	Rheita	same
Rheticus	Rhaeticus	dark area
Ricciolus Soc. I.	Riccioli	same
Riccius		same
Rocca		part of L. Aestatis
Rothmannus	Rothmann	L. Timoris
Sacrobuscus	Sacrobosco	same
Santbechius	Santbech	same
Sasserides		Ball C
Scheinerus Soc. I.	Scheiner	same
Schikardus	Schickard	same
Schillerus	Schiller	Schiller basin dk. area
Schomberger Soc. I.	'Soc. I.' dropped	same
Seleucus		same
Seneca		same
Simpelius Soc. I.	'Soc. I.' dropped	same

Name from list or map	Modern spelling	Modern feature name
Sirsalis		same
Snellius		Stevinus
Sosigenes		Julius Caesar
Stadius		dark area
Stevinus		Snellius
Stoeflerus	Stöfler	same
Stiborius		same
Sulpicius Gallus		Sulp. Gal. M brt. spot
Tannerus		Manzinus
Taruntius		same
Tatius	Tacitus	same
Thales		Strabo
Theaetetus		'bay' at Calippus omega
Thebit	(not in list)	same
Theon iun.	Theon Jun.	Taylor
Theon sen.		Taylor A
S. Theophilus Alex.	Theophilus	same
Timaeus		Plato psi etc.
Timocharis		same
Tycho	(not in list)	same
Valtherus	Walter	same
Vendelinus		same
Vernerus	Werner	same
Vietae	Vieta	Vieta+A+B+Fourier
Vitruvius		Dawes
Zenophanes		same
Zagutus	Zagut	same
*Zoroaster		M. Humboldtianum
Zucchius Soc. I.	'Soc. I.' dropped	same
Zupus		dark spot

B LARGE DARK AREAS

Name from list or map	translation (approx)
Oceanus Procellarum	Ocean of Storms
M. Crisium	Sea of Crises
M. Fecunditatis	Sea of Fruitfulness
M. Frigoris	Sea of Cold

Name from list or map	translation (approx)
M. Humorum	Sea of Moisture
M. Imbrium	Sea of Showers
M. Nectaris	Sea of Nectar
M. Nubium	Sea of Clouds
M. Serenitatis	Sea of Serenity
M. Tranquillitatis	Sea of Tranquillity
M. Vaporum	Sea of Vapours
Sinus Aestuum	Bay of Hot Days
**Sinus Epidemiarùm	Bay of Epidemics (?)
Sinus Iridum	Bay of Rainbows
Sinus Roris	Bay of Dew
†Palus Nebularum	Marsh of Fogs
†Palus Nimborum	Marsh of Rain clouds
Palus Putredinis	Marsh of Decay
Palus Somni	Marsh of Sleep
Lacus Mortis	Lake of Death
Lacus Somniorum	Lake of Dreams
†Stagnum Glaciei	Swamp of Ice

C LARGE LIGHT AREAS. NOTE THAT NONE OF THESE NAMES IS USED ON MODERN MAPS.

Name from list or map	Translation (approx)
Terra Caloris	Land of Heat
Terra Fertilitatis	Land of Fertility
Terra Grandinis	Land of Hail
Terra Mannae	Land of Manna
Terra Nivium	Land of Snows
Terra Pruinae	Land of Frost
Terra Sanitatis	Land of Healthiness
Terra Siccitatis	Land of Dryness
Terra Sterilitatis	Land of Sterility
Terra Vigoris	Land of Vigour
Terra Vitae	Land of Life
Insula Ventorum	Island of Winds
Peninsula Deliriorum	Peninsula of Insanities

Name from list or map	translation (approx)
Peninsula Fulgurum	Peninsula of Lightning
Peninsula Fulminum	Peninsula of Thunder
Littus Eclipticum	Ecliptic Shore

* These 23 names not used in modern maps.

† These 3 names are not used in modern maps; remainder are still official but with these small changes: Riccioli's Mare Nectaris now includes modern Sinus Asperitatis; his Sinus Aestuum is now Sinus Medii, while our Sinus Aestum is the dark area immediately to the north; his Palus Putredinis is now anonymous, the name having been transferred to the dark bay adjacent on the north-east; and the boundaries for Lacus Mortis and Lacus Somniorum are now a little different.

** Renamed Palus Epidemiarum by Schmidt.

APPENDIX H

SCHRÖTER'S NEW NAMES

Schröter's new names, using his spelling. This list includes four from M. Hell's list, to which he refers, also five from Van Langren and one from Allard, neither of which maps he saw. (M)=mons, montes.

Alhazen	Doerfel (M)	Lambert	Picard
Arnold	Doppelmayer	le Gentil	Pico (M)
Azout	Eimmart	Leibnitz (M)	Pingré
Bailly	Euler	Lexell	Römer
Bernoulli	Feronce	Lichtenberg	Rook (M)
Bianchini	Gärtner	Louville	Rost
Boscowich	Godin	Lubinietzky	Scharpius
Bradley (M)	Hadley (M)	Mairan	Segner
Briggs	Hase	Malvasia	Short
Cassini, J.D.	Hausen	Maraldi	Silberschlag
Cassini, J.J.	Heinsius	Maupertuis	Smith
Condorcet	Hell	Mayer, C.	Street
d'Alembert (M)	Hermann	Mayer, F.C.	Taquet
de Fontenelle	Hooke	Mayer, T.	Vitello
de Ulloa	Horrebow	Mercator	Wargentin
de la Caille	Huyghens (M)	Mont Blanc (M)	Weigel
de la Condamine	Kästner	Mylius	Wilson
de la Hire (M)	Kies	Neper	Wing
de l'Isle	Kirch (M)	Newton	Wolff (M)
Desplaces	Kraft	Palitsch	Wurzelbauer

Notes on names
Rost, Scharpius, Wolff first used by Hell.
Briggs, de Ulloa, Huyghens, Neper, Taquet first used by Van Langren.
Cassini, J.D. first used by Allard.
de Ulloa, Desplaces, Feronce, Malvasia, Mayer, F.C., Mylius, Smith, and Wing never used subsequently.
Minor spelling changes in current use: Auzout, Bernouilli, Boscovich, Cassini, Fontenelle, la Caille, la Condamine, la Hire, Delisle, Huygens, Krafft, Lubiniezky, Palitzsch, Sharp, Tacquet.

APPENDIX I

MÄDLER'S NEW NAMES

Mädler's new names, using his spelling. Includes four from Van Langren, whose map he never saw. (M)=mare.

Agatharchides	Diophantus	Laplace, Prom.	Ramsden
Airy	Drebbel	Lavoisier	Reichenbach
Ansgarius	Egede	Legendre	Repsold
Apollonius	Encke	Lehmann	Réaumur
Arago	Euclides	Letronne	Ritter
Baco	Fermat	Lindenau	Rosenberger
Baily	Flamsteed	Linné	Ross
Barrow	Fourier	Littrow	Sabine
Beaumont	Fra Mauro	Lohrmann	Saussure
Behaim	Franklin	Maclaurin	Schubert
Berzelius	Frauenhofer	Magelhaens	Schumacher
Bessel	Gambart	Marco Polo	Scoresby
Biela	Gauss	Marinus	Sömmering
Biot	Gay Lussac	Mason	Steinheil
Boguslawsky	Gérard	Messier	Strabo
Bohnenberger	Gioja	Mösting	Struve
Bonpland	Guerike	Nasireddin	Taylor
Borda	Guttemberg	Nearch	Torricelli
Bouguer	Hahn	Nicolai	Tralles
Boussingault	Hanno	Oersted	Ukert
Bouvard	Hansen	Oken	Ulugh Beigh
Buch	Hansteen	Olbers	Vasco de Gama
Burckhardt	Harding	Oriani	Vega
Bürg	Hekatäus	Pallas	Vlacq

Büsching	Herodot	Parrot	Wollaston
Carlini	Hesiodus	Parry	Zach
Cavendish	Hippalus	Pentland	
Clairaut	W. Humboldt	Piazzi	
Colombo	Jacobi	Pictet	
Cook	Jansen	Plana	Altai-Gebirg
Cuvier	Inghirami	Playfair	Australe, (M)
Damoiseau	Kant	Pons	Cordilleren-Geb.
Davy	Klaproth	Pontécoulant	Hercynii Montes
Deluc	Lagrange	Poisson	Humboldtianum, (M)
Descartes	Lapeyrouse	Polybius	Pyrenäen-Gebirg

Notes on names

Colombo, Descartes, Euclides, Vlacq first used by Van Langren.

All names used in NLF, but Oriani and Hercynii Montes not used after 1961.

Minor spelling changes in current use: Fraunhofer, Gay-Lussac, Gerard, Guericke,
    Gutenberg, Hecataeus, Herodotus, la Pérouse, Vasco da Gama, Pyrenaeus Montes.

APPENDIX J

BIRT'S AND LEE'S NEW NAMES

New names by Birt and Lee (L), who collaborated on the BA map.

Adams	Dawes	Janssen	Piton (M)
Alexander	de la Rue	Lassell	Pollock
Archidaeum Prom.	de Morgan	Lavinium Prom.	Robinson
Argelander	Delaunay	Lee	Rosse
Babbage	Donati	Livingstone	Schiaparelli
Baker	Faraday	Lockyer	Schmidt
Ball	Faye	Mädler (1)	Secchi
Beer	Foucault	Mädler (2)	Sheepshanks
Bellot	Franklin, J.	Main	Smyth, Piazzi
Birmingham	Grove	Manners	Somerville
Bond, G.P.	Gwilt, G.	Maury	South
Bond, W.C.	Gwilt, J.	McClure	Stanley
Brayley	Halley	Miller	Straight Range
Carrington	Harbinger Mts.	Moigno	Struve, Otto
Cayley	Herschel, C.	Murchison	Teneriffe Mts.
Chacornac	Herschel, J.F.W.	Olivium Prom.	Ward
Challis	Hind	Percy Mts.	Whewell
Coxwell Mts.	Horrox	Peters	Wrottesley
Crozier	Huggins	Phillips	
Daniell	Jackson–Gwilt	Photographica, Terra	
Chevallier(L)	Goldschmidt (L)	Mitchell (L)	Shuckburgh (L)
Glaisher, Mt (L)	Maclear (L)	Rümker (L)	Smythii, Mare (L)

Notes shown overleaf

Notes on names

Halley first used by Hell.

Archidaeum Prom., Baker, Coxwell Mts, Franklin, J., Gwilt, G., Gwilt, J., Jackson–Gwilt, Livingstone, Percy Mts, Terra Photographica, Pollock, Somerville, Stanley, Ward not used in NLF.

Somerville reinstated (in different location) in 1976.

Argelander, Beer, Carrington, Faraday, Jansen, Lockyer, Mädler, Peters, Rümker appear on Schmidt's map in different locations.

Mädler used on two locations.

Minor spelling change: Horrocks.

Birt also named the separate peaks of the Teneriffe Mts as follows: Petora, Guajara, Rambleta, Alta Vista, and Chajorra, but these were never used in his map or elsewhere.

APPENDIX K

NEISON'S NEW NAMES

Neison's new names are as follows.

Birt	Gruithuisen	Lubbock	Webb
Cauchy	Kinau	Newcomb	Wichmann
Clausius	Kunowsky	Nicollet	
De Vico	Lacroix	Peirce	

Birt, Kunowsky appear on Schmidt's map (in different locations).

APPENDIX L

SCHMIDT'S NEW NAMES

Schmidt's new names. Because Schmidt, Neison, and Birt and Lee were working almost concurrently, some names appear in two of the lists.

Agassiz, Prom	Demonax	Kirchhoff	Peters
Ampère (M)	Deville, Prom.	Krusenstern	Protagoras
Argelander	Dove	Kunowsky	Regnault
Asclepi	Epicurius	le Verrier	Reimarus
Barth	Epimenides	Liebig	Rümker
Beer	Faraday, Cap	Lockyer	Schwabe
Birt	Feuillé	Luther	Serao (M)
Breislak	Fresnel, Cap	Lyell	Sina
Brisbane	Galle	Mädler	Spallanzani
Bunsen	Galvani	Mallet	Timoleon
Carrington	Haidinger	Melloni	Volta
Celsius	Hamilton	Monge	Wallace
Chamisso, Cap	Heis	Möstlin	Watt
Chladni	Helmholtz	Naumann	Wöhler
Cusanus	Hencke	Neumayer	Young
d'Arrest	Heraclitus	Nöggerath	Zeno
Daguerre	Ideler	Opelt	Zöllner
Darwin	Janssen	Palmieri	
de Gasparis	Kaiser	Peirescius	
Dechen	Kane	Petermann	

Minor spelling changes in current use: Feuillée, Maestlin, Sinas.
Barth, Cap Chamisso, Epicurius, Hencke, Melloni not used subsequently, Timoleon not used after 1961.
Birt, Kunowsky appear on Neison's map (in different locations).
Argelander, Beer, Carrington, Faraday, Janssen, Lockyer, Mädler, Peters, Rümker appear on Birt's map (in different locations).
Cusanus first used by Van Langren ('Cusae'), then by Riccioli.

APPENDIX M

FRANZ'S NEW NAMES

Franz's new names are as follows.

Hall	Abel	Kelvin	Jungnitz
Aestatis (M)	Hiemis (M)	Orientale (M)	Undarum (M)
Anguis (M)	Marginis (M)	Parvum (M)	Veris (M)
Autumnis (M)	Novum (M)	Spumans (M)	(mare trans Hahn)

Jungnitz dropped, the crater being already named Demonax by Schmidt. Abel not in NFL, but reinstated in 1961 on same crater; Hall and Kelvin moved to different features in NLF; M. trans Hahn not in NLF; M. Hiemis, M. Parvum and M. Novum dropped in 1961; M. Autumnis, M. Veris, M. Aestatis changed to Lacus Autumni, Veris and Aestatis in 1970; Maria Anguis, Marginis, Orientale, Spumans, Undarum remain unchanged.

APPENDIX N

KRIEGER'S AND KÖNIG'S NEW NAMES

New names by Krieger and König are as follows.

Angström	Gould	Moltke	Seeliger
*Antural	Gyldén	*Neutra	*Semlja
Auwers	Holden	*Neutra Streifen	*Semlja Sinus
*Banat	*Jekaterinburg Damm	*Nowaja	*Sosigenes Bucht
*Banat Sinus	*Jekaterinburg Pass	Oppolzer	Spörer
**Banat, Kap	*Karische Strasse	Palisa	Suess
Bruce, Miss	*Keeler	Pickering, E.	*Tatra
Burnham	Kelvin, Kap	Pickering, W.	Tempel
*Cisneutra	Klein	*Pietrosul	Tisserand
*Cissemlja	Lade, von	**Pietrosul Bucht	*Transsemlja
Daguerre	Lamont	Prinz	**Ural
Dembowsky	Lepaute	Puiseux	Vogel
Draper	Lick	**Riphaeus Boreus	Weiss
Dunthorne	Lippershey	**Riphaeus Major	Williams
Elger	Loewy	**Riphaeus Medius	Wolf, Max
Franz	Lohse	**Riphaeus Minor	Yerkes
Gaudibert	Marth	Ritchey	
**Gay-Lussac Sinus	*Matra	**Schneckenberg	

* Not used in NLF.
** Not official after 1961.
Minor spelling changes in current use: Dembowski, Kelvin Prom., Lade Bruce, Pickering, Wolf; Keeler reinstated in different location, 1970.

APPENDIX O

FAUTH'S AND DEBES'S NEW NAMES

Fauth's and Debes's new names are as follows.

Fauth's: Brenner, Lagalla, Montanari, Trouvelot, Weinek, Drygalski.

Schupmann, Hörbiger, Mare Horologii, Mare Avis, Regio Paupertatis, Regio Sanitatis, Sven Hedin not used in NLF (some of these may post-date the publication of NLF). Hedin reinstated on same feature in 1964.

Debes's: Jura.

APPENDIX P

LAMÈCH'S NEW NAMES

Lamèch's new names are as follows.

Athinagoras	Dominique	Moreaux	Roy
Baillaud	Drossos	Moumouris	Rudaux
Baldet	Dupont	Müller	Schlumberger
Bigourdan	Fredericos	Myriame	Siredey
Blagg	Gasser	Neison	Soulayou
Bojena	Glasenapp	Pasteur	Stephanides
Boquet	Hellène	Piérot	Touchet
Danjon	Kephalinos	Plakidis	Turner
Darney	Klepesta	Potamos	Voutzinas
Delmotte	Levisky	Rodés	Wally
Deseilligny	Mariane	Rossard	Zinger

Of these, only Baillaud, Blagg, Darney, Delmotte, Deseilligny, Müller, Neison, and Turner used in NLF, 1935.
Baldet, Danjou, Glasenapp (Glazenap), Pasteur, and Zinger (Tsinger) reinstated (in different locations) in 1970.

OTHER NEW NAMES IN NAMED LUNAR FORMATIONS

Other new names used in NLF, 1935, are as follows.

By Blagg: Spitzbergen ('pointed peaks', and from shape of this mountain group).

By Müller: Andĕl, Debes, Fauth, König, Lamèch, Saunder, Wilkins.

By IAU Commission 17: Brown, Proctor.

By Peucker: da Vinci.

By Wilkins: Mee, Goodacre.

Notes

Alpine Valley and Straight Wall. These names have been attributed to Elger, but Neison (1874) lists the former, and Birt has 'Straight Wall' on his general map (1870s).

Schröter's Valley. Name attributed to W. Pickering in NLF.

All names have been retained.

APPENDIX R

WILKINS'S NEW NAMES

New names in Wilkins's 300-inch Moon map. None accepted at the 1948, 1952 or 1955 General Assemblies of the IAU. The names had been proposed by Wilkins, Paluzíe, Moore, Mee, McDonald, Arthur, Cameron, Hoag, Adams, and Nicholson. Of these 96, 16 have been used subsequently, all but a few in different locations.

Abineri	Dublier	Krosigk	Reese
Aller	Dyson, Mt.	Landerer	Renart
Alter	Eddington	La Paz	Reypastor
Amundsen	Einstein	Lenham	Rhodes
Antoniadi	Emley	Liddiard	Rodés
Armenter	Esquivel	Lowe	Romaña
Arthur	Febrer	Lower	Russell
Aymat	Fisher	Lyot	Sacco
Ball, L.F.	Fresa	MacDonald	Saheki
Barange	Gant	Millás	Santacruz
Barcroft	Garcia-Gomez	Moore	Scott
Barker	Giner	Najerá	Shackleton
Bartlett	Graham	Nansen	Sisebuto
Baum	Green	Novellas	Smith
Benitez	Haas	O'Kell	Steavenson
Bertaud	Hallowes	O'Neill	Thornton
Bolton	Hare	Orús	Trewman
Burrell	Harris	Paluzíe	Väisälä
Buss	Hauet	Peary	Vernet
Caramuel	Hill	Polit	Virgil
Clarkson	Ibañez	Porter	Wagner
Comas Sola	Incognito, Mare	Porthouse	Watts
Cooke	Ingalls	Pratdesaba	Whipple, Mt.
Cortés	Jiyah	Raurich	Whitaker
de Bergerac	Juán	Recorde	Wright

APPENDIX S

IAU LUNAR NOMENCLATURE RESOLUTIONS, 1961

Resolutions regarding lunar nomenclature approved at the IAU General Assembly, 1961, and published in *Tr.IAU*, Vol.XIB, 1962.

RESOLUTION NO. 1

'For compiling new maps of the Moon, the following conventions are recommended:

(*a*) *Astronomical* maps for purpose of telescopic observations are oriented according to the astronomical practice, the South being up. To remove confusion, the terms East and West are deleted.

(*b*) *Astronomical* maps, for direct exploration purposes, are printed in agreement with ordinary terrestrial mapping, North being up, East at right and West at left.

(*c*) Altitudes and distances are given in the Metric System.'

RESOLUTION NO. 2

1. 'For designating the lunar surface features, it is recommended that the previous rules be followed, revised and improved as follows:

(*a*) Craters and rings, or walled plains, are designated by the name of an astronomer or prominent scientist *deceased*, written in the Latin alphabet, and spelt according to the recommendation by the country of origin of the scientist named.

(*b*) Mountain-like chains are designated in Latin by denominations allied with our terrestrial geography. Names are associated with the substantive *Mons* according to the Latin declination rules and spelling. (Three exceptions, Mons d'Alembert, Mons Harbinger and Mons Leibnitz, are preserved due to long usage).

(*c*) Large dark areas are designated in Latin denominations calling up psychic states of minds. These names are associated, according to the Latin declination rules and spelling, to one of the appropriate substantives: *Oceanus, Mare, Lacus, Palus* or *Sinus*. (The exceptions, Mare Humboldianum and Mare Smythii, are preserved, due to long usage).

231

(d) Isolated peaks are designated according to the same rules as for the craters, as well as promontories, the latter being preceded by the Latin substantive *Promontorium*. (Example: Promontorium Laplace).

(e) Rifts and valleys take the names of the nearest designated crater, preceded by the Latin substantives *Rima* and *Vallis*. (The exception Vallis Schröter is preserved).

(f) Undenominated features can be designated by their co-ordinates. They can equally be designated according to the former classical system, by taking the name of the nearest crater, followed by an upper case letter of the latin alphabet for craters, depressions and valleys, by a lower case letter of the greek alphabet for hills, elevations and peaks, and by a roman number followed by the latter r (Ir, IIr, IIIr, etc.) for the clefts.'

2. 'Accordingly, for the designation of the lunar surface features observable from the Earth, it is recommended that the International Astronomical Union Nomenclature (published in *Named Lunar Formations* by M. A. Blagg and K. Müller, London, 1935), as corrected in table III of the *Photographic Lunar Atlas* (Editor G. P. Kuiper, University of Chicago Press, 1960), be adopted, that new names be avoided, and that the following orthographic corrections be applied:

Condamine must be printed La Condamine

Lacaille	„	„	„	La Caille
Lahire	„	„	„	La Hire
Lapeyrouse	„	„	„	La Pérouse
Legentil	„	„	„	Le Gentil
Lemonnier	„	„	„	Le Monnier
Leverrier	„	„	„	Le Verrier
Régnault	„	„	„	Regnault

3. 'For the designation of the surface features on the reverse side of the Moon, it is recommended that the nomenclature reported in the *Atlas of the Far Side of the Moon* (Editors: N. P. Barbashov. A. A. Mikhailov and Y. N. Lipsky, Moscow 1960) be adopted, expressed in terms of the following table and the accompanying chart'.

Edison	Jules Verne	Mendeleev	Sovietici, Montes
Giordano Bruno	Kurchatov	Moscoviense, Mare	Tsiolkovsky
Hertz	Lobachevsky	Pasteur	Tsu Chung-Chi
Ingenii, Mare	Lomonosov	Popov	
Joliot Curie	Maxwell	Sklodowska Curie	

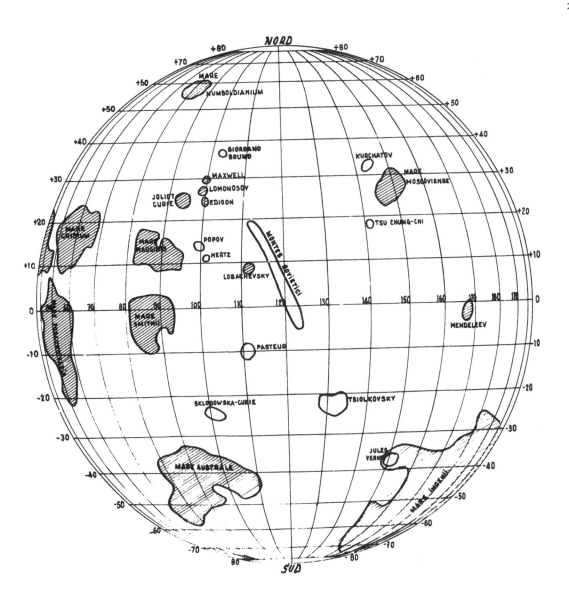

APPENDIX T

NEW NAMES IN THE *RECTIFIED LUNAR ATLAS*

New and reinstated names by Arthur and Whitaker as published in the *Rectified Lunar Atlas* (RLA, 1963) and approved by the IAU in 1964.

1 NEW NAMES

Aston	Cannon	Hayn	Pascal
Baade	Cremona	Hermite	Planck
Balboa	Dalton	Hubble	Poncelet
Balmer	Desargues	Jansky	Rayleigh
Banachiewicz	De Sitter	Jeans	Riemann
Barnard	Dubiago	Kapteyn	Röntgen
Belkovich	Fermi	Krasnov	Schlüter
Bohr	Gibbs	Lamarck	Shaler
Boltzmann	Gilbert	Lamé	Stefan
Boole	Gill	Langley	Stokes
Boss	Goddard	Liapunov	Sylvester
Brianchon	Hale	Markov	
Byrd	Hartwig	Moseley	

2 RESTORED NAME AND LOCATION

Abel (from Franz)
Drygalski (from Fauth)
Scott (from Arthur/Wilkins)

Amundsen (from Arthur/Wilkins)
Joliot–Curie (from *Luna 3* chart)

3 RESTORED NAME BUT CHANGED LOCATION

Bunsen (from Schmidt)
Einstein (from Wilkins)
Hedin (from Fauth)
Nansen (from Moore/Wilkins)
Russell (from Wilkins)
Wright (from Wilkins)

Eddington (from Cameron/Wilkins)
Hamilton (from Schmidt)
Lyot (from Wilkins)
Peary (from Moore/Wilkins)
Volta (from Schmidt)

4 CHANGES TO NLF OTHER THAN SPELLING OR LATINISATION, AS OF 1967. CHANGES IN LETTER DESIGNATIONS NOT INCLUDED HERE.

Deletions. Oriani, Timoleon, Riphaeus Boreus, Riphaeus Major, Riphaeus Medius, Riphaeus Minor, Ural, Gay-Lussac Sinus, Pietrosul Bay, Palus Nebularum, Mare Aestatis, Mare Autumni, Mare Hiemis, Mare Novum, Mare Parvum, Mare Veris, Schneckenberg, J.J. Cassini (name combined with Cassini, vague feature left anonymous), W. Pickering (name combined with Pickering, crater 'Messier A' again), Prom. Banat, Montes Hercynii, Struve (dark patch, name combined with NLF 'Otto Struve', which was shortened to 'Struve').

Moves. Names of formations for which the NLF map is not explicit; identifications those of original authority: Brisbane, Montes Cordillera, Galvani, Hausen, Le Gentil, Marco Polo, Pingré, Regnault, Seneca, Ulugh Beigh, Wolff.

APPENDIX U

ADDITIONS TO THE *NASA CATALOGUE OF LUNAR NOMENCLATURE*, RP 1097

Name	Previous name/desig.	Name	Previous name/desig.
Andersson	anon.	Lallemand	Kopff A
Ashbrook	Drygalski Q	McAuliffe	Borman Y
Beals	Riemann A	McNair	Borman A
Blanchard	Arrhenius P	Murakami	Mariotte Y
Bondarenko	Patsaev G	Nobile	Scott A
Chadwick	De Roy X	Onizuka	anon.
Chalonge	Lewis R	Pilâtre	Hausen B
Chappe	Hausen A	Resnik	Borman X
Couder	Maunder Z	Rosseland	Roche U
Faustini	anon.	Scobee	Barringer L
Florenskii	Vernadskii B	Shackleton	anon.
Fryxell	Golitsyn B	Shuleikin	Pettit T
Glushko	Olbers A	Smith	Barringer M
Hédervári	Amundsen	Sternfeld	Lodygin B
Heyrovsky	Drude S	Urey	Rayleigh A
Il'in	Hohmann T	Vinogradov	Euler P
Jarvis	Borman Z	Von Braun	Lavoisier D
Konoplev	Ellerman Q	Yakovkin	Pingré A
Kramarov	Lenz K		

APPENDIX V

....

SELECTED BIBLIOGRAPHY AND REFERENCES

Enough material has been published about the Moon to fill a medium-sized library. These references and suggestions for 'further reading' do not pretend to cover more than a small part of the general subject matter of this book, except perhaps for references dealing with the standardization of lunar nomenclature, which is sufficiently complete for anyone who wishes to follow up that subject. The remainder is set out by chapters as far as possible.

CHAPTER 1

J. Classen, Die Älteste Mondkarte, *Die Sterne*, Jg. **22**, 1 (1942), with additional material (1963)

R.E. Guiley, *Moonscapes – A Celebration of Lunar Astronomy, Magic, Legend and Lore* (New York, 1991)

T. Harley, *Moon Lore* (London, 1885, reprinted Rutland, VT, 1970)

T. Harley, *Lunar Science* (London, 1886)

J.C. Houzeau, Ce qu'on voit dans la Lune, in J.C. Houzeau and A. Lancaster, *Biblio Générale D'Astronomie*, 1 (Brussels, 1884; reprinted London, 1964), 40–3

S.L. Montgomery, *The Scientific Voice* (New York, 1996)

F.K. Pizor and T.A. Comp, (eds.), *The Man in the Moone and Other Lunar Fantasies* (New York, 1971)

J.W. Stein, S.J., Saint Albert le Grand et l'Astronomie, *Specola Vaticana Miscellanea Astronomica* **3**, No. 102, 81–8

P.J. Stooke, Neolithic Lunar Maps at Knowth and Baltinglass, Ireland. *Journal for the History of Astronomy*, **25**, 1 (1994), 39–55

C.R. Wicke, The Mesoamerican Rabbit in the Moon: An Influence from Han China? *Archaeoastronomy*, VII(1–4) (1984), 46–55

H. Wright *et al.* (eds.), *To the Moon!* (New York, 1968)

CHAPTER 2

A. Favaro, (ed.), *Le Opere di Galileo Galilei*, National Edition (Florence, 1890–1909), iii, part 1, 48

A. van Helden, *Sidereus Nuncius, or The Sidereal Messenger (of) Galileo Galilei* (Chicago 1989)

P. Humbert, La Première Carte de la Lune. *Revue des Questions Scientifiques*, **20** (Brussels, 1931), 193–204

E. Rosen, *Kepler's Conversation with Galileo's Sidereal Messenger* (New York, 1965)

O. Van de Vyver, Original Sources of some Early Lunar Maps. *Journal for the History of Astronomy*, **2** (1971), 86–97

E.A. Whitaker, Galileo's Lunar Observations and the Dating of the Composition of *Sidereus Nuncius*. *Journal for the History of Astronomy*, **9** (1978), 155–69

CHAPTER 3

Gachard, Lettre de Philip IV à l'infante Isabelle, touchant certaines luminaires découverts au ciel par Michel-Florentius Van Langren. *Bulletins de l'Academie Royale de Belgique*, **T12**, II (1845), 261–2

O. van de Vyver, S.J., Lettres de J.Ch. della Faille S.J., cosmographe du roi à Madrid, à M.-F. Van Langren, cosmographe du roi à Brussels, 1634–1645. *Archivum Historicum Societatis Iesu*, **XLVI** (1977), 73–183

CHAPTER 4

J. Hevelius, *Selenographia*, Danzig (1647), reprint (New York, 1967)

G.D. Riccioli, De Luna. *Almagestum Novum*, **1** (Bologna, 1651)

CHAPTER 5

F. Bianchini, *Hesperi et Phosphori Nova Phenomena* (Rome?, 1727)

J.-D. Cassini, Avertissement touchant l'Observation de l'Éclipse de Lune, qui doit arriver la nuit du 28 Juillet prochain. *Memoires de l'Academic Royale des Sciences*, X (Paris, 1692), 126–9

E.G. Forbes, The Life and Work of Tobias Mayer (1723–1762). *Quarterly Journal of the Royal Astronomical Society*, **8** (1967), 227–51

E.G. Forbes, *Tobias Mayer's Opera Inedita: the First Translation of the Lichtenberg edition of 1775* (London, 1971)

R. Hooke, *Micrographia*, (London, 1665; reprinted New York, 1961)

C. Huygens, *Oeuvres Completes*, **15** (Amsterdam, 1967) 155, 158, 160

Cherubin d'Orléans, *Dioptrique Oculaire* (1671)

G. Montanari, *Ephemerides Novissimae . . .* (Modena, 1662, for Montanari's map)

T. Weimer, Carte de la Lune de J.-D. Cassini. *The Moon and the Planets*, **20** (1979), 163–7

J. Zahn, *Speculum Physico-mathematica Historicum* (Nüremburg, 1694), for Eimmart's lunar map

CHAPTER 6

W.F. Ryan, John Russell, R.A., and Early Lunar Mapping. *Smithsonian Journal of History*, **1** (1966), 27–48

J.H. Schröter, *Selenotopographische Fragmente*, **1** (Lilienthal, 1791) and **2** (Göttingen, 1802)

CHAPTER 7

W. Beer and J.H. Mädler, *Mappa Selenographica* (Berlin, 1834 and 1877)

W. Beer and J.H. Mädler, *Der Mond* (Berlin, 1837)

W.G. Lohrmann, *Topographie der Sichtbaren*

Mondoberflaeche (Leipzig, 1824)

W.G. Lohrmann, *Mondcharte in 25 Sectionen, herausgegeben von J. Schmidt* (Leipzig, 1878)

CHAPTER 8

T.G. Elger, *The Moon* (London, 1895)

W. Goodacre, *Map of the Moon* (London, 1910)

W. Goodacre, *The Moon* (Bournemouth, 1931)

J. Nasmyth and J. Carpenter, *The Moon* (London, 1874)

E. Neison, *The Moon* (London, 1876)

J.F.J. Schmidt, *Charte de Gebirge des Mondes* (Berlin, 1878), book and map

CHAPTER 9–11

Yu I. Efremov *et al.* (eds.), *Atlas Obratnoi Storony Luny*, II (Moscow, 1967)

Yu I. Efremov *et al.* (eds.), *Atlas Obratnoi Storony Luny*, III (Moscow, 1975)

P. Fauth, *Mond Atlas* (Bremen, 1964 – with his *Mondatlas*, 1932)

R. König, *Joh. Nep. Krieger's Mond-Atlas*, **1** and **2** (Vienna, 1912)

J.N. Krieger, *Mond-Atlas* (Trieste, 1898)

Yu N. Lipsky *et al.*, *Atlas of the Moon's Far Side*, I (translation by R.B. Rodman, New York, 1961)

H.P. Wilkins and P.A. Moore, *The Moon* (London, 1955)

The following are references to materials dealing with the standardization of lunar nomenclature, starting with the position catalogues by Franz that triggered the alerts by S.A. Saunder (chapters 9–11).

L.E. Andersson and E.A. Whitaker, *NASA Catalogue of Lunar Nomenclature*, (NASA Reference Publication 1097), (Washington, 1982)

D.W. Arthur and A.P. Agnieray, *Lunar Designations and Positions*, (Tucson, 1966) – maps from previous reference combined into four quadrants, with shading added in mare areas

D.W. Arthur *et al.*, *The System of Lunar Craters* (Tucson, 1963,4,5,6)

R.M. Batson and J.F. Russell (eds.), *Gazetteer of Planetary Nomenclature 1994* (USGS Bulletin 2129), (Washington, 1995)

M.A. Blagg, *Collated List of Lunar Formations* (Edinburgh, 1913)

M.A. Blagg and K. Muller, *Named Lunar Formations*, **1**, Catalogue; **2**, Maps (London, 1935)

J. Franz, Die Randlandschaften des Mondes. *Nova Acta. Abh. der Kaiserl Leop.-Carol. Deutschen Akademie der Naturforscher*, IC, 1 (1913), 1–95

J. Franz, Ortsbestimmung von 150 Mondkratern, *Mitteilungen der Königlichen Universitäts-Sternwarte zu Breslau*, 1 (1901), 1–70; *ibid.*, **2** (1901), 31–50

G.P. Kuiper (ed.), *Photographic Lunar Atlas*, Table 3 (Chicago, 1960)

D.H. Menzel *et al.*, Report on Lunar Nomenclature. *Space Science Reviews*, **12** (1971), 136–86

S.A. Saunder, Lunar Nomenclature,. *Memoirs of the Royal Astronomical Society*, **57** (1905), 47–8

S.A. Saunder, On the Present State of Lunar Nomenclature. *Monthly Notices of the Royal Astronomical Society*, LXVI No.2 (1905), 41–6

S.A. Saunder, Notes on Lunar Nomenclature, unpublished leaflet for presentation to General Assembly of the International Association of Academies (Vienna, May 1970), pp. 1–4

H.H. Turner, Lunar Nomenclature. *MNRAS*, LXVIII No.2 (1907), 134–45

H.H. Turner, Lunar Nomenclature. *MNRAS*, LXIX No.1 (1908), 3–7

E.A. Whitaker *et al.*, *Rectified Lunar Atlas* (Tucson, 1963)

The following all refer to the *Transactions of the International Union*, the quoted year being that of the General Assembly, not necessarily of the date of publication.

I,	(1922), 52–4, 164, 210	VIII,	(1952), 216,
II,	(1925), 54–5, 189–90, 233	IX,	(1955), 263
III,	(1928), 111–19, 240, 303	X,	(1958), nothing
IV,	(1932), 96, 237–8, 284	XI(B),	(1961), 234–238
V,	(1935), 108, 304, 372	XII(B),	(1964), 202–205
VI,	(1938), nothing	XIII(A),	(1967), 344, 345
VII,	(1948), 63, 160, 166, 169	XIII(B),	(1967), 43, 103–5

XIV(A),	(1970), 169–70		**XVIII**(B),	(1982), 331
XIV(B),	(1970), 63, 138–9, 142		**XIX**(B),	(1985), 340–1
XV(A),	(1973), 204–6		**XX**(A),	(1988), 703–4
XV(B),	(1973), 110–14, 207–17		**XXI**(A),	(1991), 613
XVI(B)	(1976), 321–69 (various sections)		**XXII**(A),	(1994), 603
XVII(B),	(1979), 285–90			

GENERAL REFERENCES

W.B. Ashworth Jr., *The Face of the Moon, Galileo to Apollo* (Catalog of exhibition, Linda Hall Library, Kanses City, MO, 1989)

E.E. Both, *A History of Lunar Studies* (Buffalo, 1962 app.)

R. Greeley and R.M. Batson (eds.), *Planetary Mapping* (Cambridge, 1990)

A.J. Kinder, *Who's Who in the Moon* (in preparation, an update of the *First Memoir of the Historical section of the British Astronomical Association*, 1938)

Z. Kopal and R.W. Carder, *Mapping of the Moon, Past and Present* (Dordrecht, 1974)

P. Maffei, *Carte Lunari di Ieri e di Oggi* (Florence, 1963)

A. Rükl, *Hamlyn Atlas of the Moon* (London, 1991)

O. van de Vyver, Lunar Maps of the XIIth Century. *Vatican Observatory Publications*, **1** (1971), 71–114

E.A. Whitaker, Selenography in the 17th Century. *Planetary Astronomy from the Renaissance to the Rise of Astrophysics*, 2A (Cambridge, 1989)

D.E. Wilhelms, *To a Rocky Moon, a Geologist's History of Lunar Exploration* (Tucson, 1993)

INDEX